ROOTIN' in the RHIZOSPHERE:
Growing up in Ecosystem Ecology

David C. Coleman

Odum School of Ecology
University of Georgia
Athens, GA 30602-2602

Bilbo Books Publishing
ATHENS, GEORGIA

ROOTIN' IN THE RHIZOSPHERE:
Growing up in Ecosystem Ecology

Copyright © 2022 by David C. Coleman, *davec@uga.edu*

All rights reserved. No part of this book may be used or reproduced by any means, graphic, electronic or mechanical, including photocopying, recording, tape or by any information storage retrieval system without the written permission of the publisher, except in the case of brief quotations embodied in critical articles and reviews.

All photos published in this book belong solely to the author, except for the photos where indicated, which belong to the University of Georgia Odum School of Ecology and D.A. Crossley Jr.

ISBN: 978-1-7364598-9-8
Printed in the United States of America

Published in the United States of America by
Bilbo Books Publishing in Athens, Georgia.
www.BilboBooks.com
bilbobookspublishing@gmail.com
(706) 549-1597

COVER DESIGN BY GREGORY BOROKOWSKI
INTERIOR DESIGN BY TRACY N. COLEY

Contents

Prologue 7

PART ONE: Early History, Education through Post-doctoral Years
1. An early leap into the future 10
2. My musical background 14
3. Early years back East 17
4. Midsummer trip to Montana in 1944 23
5. Big change, move to San Diego, California 29
6. The summer voyage of "Big Sister" 32
7. The early years in San Diego 36
8. Major setback for Frank Coleman 39
9. The late 1940's in San Diego 43
10. Musical influences 44
11. Activities in Junior High school 50
12. My solo trip to Montana 53
13. Domestic help 55
14. North to South by rail in the western US 57
15. International influences 60
16. Origins of my life-long interests in Biology 61
17. My grad school years at the U. of Oregon 68
18. The supportive nature of my parents 72
19. Post-doctoral Experiences in the UK 73
20. Further courtship 80
21. Early years as an academic researcher 81
22. Major change in the Coleman family 88
23. A retrospective look at my years at SREL 89
24. U.S. International Biological Program (IBP) 90
25. Joining the IBP/Grasslands Biome Program 95
26. International scientific experiences 101
27. Post-IBP Research Experiences 107

PART TWO: Later Professional Experiences

28. Post-IBP Travel 110
29. Scientific travel experiences in the 1970's 117
30. A sabbatical in New Zealand 119
31. Return to Colorado State University 122
32. Research, teaching and travel 123
33. Speech Difficulties in my Life 126
34. Interviewing for UGA 128
35. Institute of Ecology, UGA 136
36. Repercussions from leaving CSU 141
37. Tropical Ecology: Studies in Kenya 142
38. Studies in collaborative research 148
39. The Horseshoe Bend experiments 150
40. Editorial experiences 153
41. East Asia and soil ecology 154
42. Long-Term Ecological Research Network (LTER) 156
43. LTER Site Reviews, a great learning experience 160
44. Experience with NSF Proposal Reviews 161
45. Graduate student advising experience 162
46. Ecological research projects in the American tropics 166
47. Early phases of retirement, 2006 onward 170
48. Major synthesis publications 174
49. My viewpoints on Global Climate Change 175
50. Persons and personalities who influenced me 176

The World Travels of Dave Coleman 178

References 181

"David Coleman's life experiences mirror a journey running on eternally connected tracks through a myriad of ecosystems, reflecting his personal and professional achievements. The story clearly depicts his journey through corridors-labs/countries/continents over more than four decades truly reflects the path many soil organisms take in the maze of soil pore networks..."

> — Gupta Vadakattu, Senior Research Scientist, Commonwealth Scientific and Industrial Research Organization, Glen Osmond, South Australia

"All that research in exotic places! Quite a career."

> — Dac Crossley, Distinguished Research Professor of Ecology, Odum School of Ecology, University of Georgia

Prologue

At numerous times my life as an ecologist has consisted of playing a supporting role, in the activities of many research groups in different places and times. I managed to originate a few ideas over the decades, including studies of the **rhizosphere**, or root-influenced, processes in a wide range of agricultural, grassland and forested soils. I entitled this memoir *Rootin' in the Rhizosphere* because much of my research was belowground in soil, and the root-influenced zone, the few mm. of soil microbes and fauna affected by the roots, was one of the key areas of research for my graduate students and me. The cover picture depicts a one square cm. cross-section of a grassland or agricultural soil, untilled, with a root and mycorrhiza on the outer right edge, and various soil micro-fauna, including nematodes and mites, inhabiting soil pores.

Rootin' in the Rhizosphere was a song that a few graduate students sang about me and our research group at the Horseshoe Bend research area of the Institute of Ecology at the University of Georgia.

These memoirs cover a journey over the past half century, providing you, the reader, a front row seat to observe several of the important developments to arise in the relatively new area of Ecosystem studies in the time span from 1965 to the present (Coleman, 2010).

The book is in two major parts: 1) formative years and education through graduate school and two post-doctorates, and 2) later years as my career unfolded over nearly fifty years.

My story begins with a signal or seminal event at age 10 that was a harbinger of many future developments.

David Coleman and dog Tippy, New York 1947

PART ONE:
Early History, Education through Post-doctoral Years

1.
An early leap into the future

It was an afternoon like any other one in the spring of 1949 in Southern California. My friends and I were restless, and decided to look around the area of Sunset Cliffs on Point Loma, southwest of my home on Granger Street, one mile south of the sleepy town of Ocean Beach, a suburb of San Diego. We scouted out some of the areas to the south, just north of the housing development of Azure Vista. The development had little duplexes built at the peak of wartime activity when the Consolidated Vultee (later Convair) corporation turned out B-24 bombers by the dozen every week in 1943 and 1944.

We were interested in exploring a fenced-off area that had an open hole above a large cave in the cliffs. As boys will, we began daring each other to somehow gain access and see what was in the big cave-like area. We needed lots of rope to let one of us down there. In our garage at home was a large skein of 200 or more feet of nylon rope, from a cross-country trip my dad and I made two years earlier (described later). I walked home, got the rope and then returned with it. Michael, David and Toby were waiting for me and asked if I was volunteering to be lowered down there. Being impetuous and usually not the most daring of our gang of four or so, I said sure, and began tying the rope around my middle. The others began wrapping rope around one of the big white posts holding the fence up, and we seemed ready for action.

What I did next was a complete change in my life, one that surprised the boys and me. I assumed that we were ready, when the guys were nowhere near prepared to hold onto me, or belay me down. I inched toward the large opening, eyeing the waves coming in halfway up on the sand inside, and told them I was going in thirty seconds or less. They said: "Wait a bit; we aren't ready yet."

I said I was ready and told them to hold on. That is the last thing I recall before suddenly accelerating through the air, rope flying out above me, landing with a sudden thump on the soft, wet sand more than forty feet below. The impact forced all the air out of my lungs, and I was unable to breathe for nearly a full minute. Then I yelled and cried, and felt a great pain in my left side, where I had landed. The guys yelled that they were sending for help, and a long afternoon's wait ensued. I wondered what would happen next? Would I be badly injured? I was oblivious to the tidal schedule, and hoped for the best.

My mother and grandparents were taking my two-year old sister Margery for a walk in a carriage along the cliffs nearly one hour later, and they asked bystanders what had happened. By then, fire department and police vehicles had arrived, and a frogman was sent in from the ocean side to check me out. After another hour, a crane arrived from nearby Mission Beach, and they lowered a basket into the hole and lifted me out of there. Grandfather Coleman rode with me the ten miles to Mercy Hospital in downtown San Diego. As it was not a life-threatening emergency, fortunately, no siren sounded.

One vignette from this adventure stands out now. During the entire trip across town, which took close to forty minutes or more, my grandfather sat there, keeping a benevolent eye on me, but at no time commenting on the situation or asking me how I was. As with older relatives in general, silence was golden in those days. I might have been more worried if Grandfather had expressed his concerns. At the emergency room I was X-rayed, and found to have severe bruising on my left hip. I went home after an overnight stay.

My father was not there, as he was in a hospital near San Bernardino, recovering from a bout of bipolar disorder. My mother and grandparents seemed to take it in stride, but that was back in the days when people kept a stiff upper lip whenever a major mishap had occurred.

The accident marked a turning point in my life. I determined to get a newspaper route, be independent of the neighborhood boys, and learn to explore, but with less impetuosity and better planning.

In retrospect, my luck was amazingly good. The tide was going out, and I was never in danger of being drowned. This is all the more remarkable as none of us boys were aware of where in the tidal cycle we were on that day in spring of 1949. My fall was only a few weeks after the tragedy of a young girl, Kathy Fiscus, who had a fatal fall down a well in central California. Her fall galvanized the entire country. My accident made the front pages of the California papers, and even led to inquiries from a few people back east in New York state, from where we had moved only two years earlier.

In the fifty years of my professional life, I have met many notable people. My interactions with them were mostly enjoyable and always memorable. I entered into ecosystem science via the back door, with an eventful ten months' postdoctoral (in 1964-65) with Amyan Macfadyen at the University College of Swansea, Glamorganshire, U.K., followed by a six-year sojourn at the Savannah River Ecology Laboratory. I then moved on into large projects funded by the Ecosystem Studies program of the National Science Foundation (NSF), and later into several academic appointments at Colorado State University and the University of Georgia. The trajectory I followed seems almost straightforward now, but was full of uncertainties and unknowns every step of the way.

At numerous times my life as an ecologist has consisted of playing a supporting role in the activities of many research groups in different places and times. I managed to originate a few ideas over the decades, including studies of rhizosphere, or root-influenced processes in a wide range of agricultural, grassland and forested soils. I entitle this memoir *Rootin' in the Rhizosphere*, because much of my research was belowground in soil and the root-influenced zone (the few millimeters of soil immediately affected by the roots was one of the key areas of research for my graduate students, postdoctorals and me).

Rootin' in the Rhizosphere was a song that a few graduate students sang about me and our research group ("Old Dave Coleman rootin' in the Rhizosphere…") at the Horseshoe Bend research area of the Institute of Ecology at the University of Georgia.

Over the past half century, I have been privileged to have a front row seat to observe several of the important developments in the relatively new area of Ecosystem studies in the time span from 1965 to the present (Coleman, 2010).

2.
My musical background

I grew up in a musical family, with my mother, Alice Cowan Coleman, being an accomplished pianist and singer (soprano) in her own concerts in the Matinee Musicale in Beacon, New York, during the World War Two years. Later on, she and my father sang in madrigal and choral groups in the San Diego area. She encouraged me to play clarinet in fifth grade, and then bassoon from the sixth grade onward. My sister was similarly encouraged to play the flute. In 1954, my parents generously ordered a bassoon for me, custom-made in the Heckel Company's factory in Biebrich, West Germany (as it was then known). I played this instrument in the Youth Symphony that met weekly in Balboa Park in central San Diego. I was thus primed to make the most of the musical milieu at Reed College.

Reflecting the uncertain times during the Great Depression, my parents moved several times in the first few years of their marriage. My father, Francis Coleman, met my mother, Alice Cowan, at Reed College, in Portland, Oregon in 1928, when he was a senior, majoring in Physics, and my mother, a freshman, needed tutoring in Physics (Figures). Reading between the lines in my mother's daily diary from those years, she preferred to play the field, and date several young men, whereas my father-to-be was very smitten with her. Graduating in 1929, Francis went on to pursue a Master's degree in Physics at the University of California, Berkeley, getting that degree in 1931. He then applied for and was awarded a Rhodes Scholarship to attend Oxford University between 1931 and 1933, obtaining a D. Phil. Oxon. (Doctoris Philosophiae Oxoniensis) in Physics. One of the "perks" of his degree was a cardinal red gown decorated with blue cloth panels, and a red and blue cap. What a contrast to the usual black graduation caps and gowns!

Alice attended Reed during her sophomore year as well. With family funds being very scarce in the early stages of the Great Depression, her father urged her to attend the University of Montana for her junior year. She got her B.A. degree in English literature at the University of Washington in Seattle in the spring of 1932. On occasional visits back to the USA, Francis visited Alice and continued their courtship. Francis proceeded to Cambridge University for postdoctoral research, and Alice joined him there in the summer of 1933 to travel around the U.K. and also in Germany. They set a wedding date for early August 1934, and Alice returned to the USA to teach grade-school students at the Rocky Boy Indian Reservation a few miles from her home in Box Elder, in north-central Montana. After a honeymoon trip to Glacier National Park, Frank and Alice traveled to Albuquerque, New Mexico, where Francis taught introductory Physics at the University of New Mexico for two years. By my Dad's proud account, in the year he was hired (1934) his job was one of only three faculty openings in college or university physics departments nationwide.

In the fall of

Francis F. Coleman and Alice Cowan, 1932

1936, my parents-to-be moved to Bennington College in Bennington, Vermont, where my father taught physics and seminars in modern topics in arts and science. Francis, a rather sensitive and quiet sort, became quite upset at the political developments in Germany. Frank and Alice traveled to Germany in the summer of 1937 and visited friends near Darmstadt, in southern Germany. They were concerned about the moves on Austria and Czechoslovakia threatened by the Hitler government.

3.
Early years back East

I was born on November 7, 1938, in Bennington. By that time, my dad had what was termed a "nervous breakdown" (called bipolar condition now), and had entered a sanitarium in Hartford, Connecticut in the early fall of 1938. My mother told me many years later that he didn't even see me until I was ten months old in September of 1939. Dad received many insulin and electric shock treatments and took a long time to recover. Dad fought off the effects of morphine used as part of the treatments and it worried my mother considerably.

My parents then moved to Long Island, where Francis taught high school science courses at a private school. This was a temporary stopgap until my dad had recovered further. The Texas Company offered him a research physicist job in early 1941 at their large laboratory in Glenham, near Beacon, New York, My folks moved to Beacon for a few months and then to Fishkill, five miles east of the Hudson River.

Our home from 1941-1947 in Fishkill was an old Victorian-era house. Fishkill (Dutch for "fish creek") was a sleepy town in Dutchess County, ca. sixty miles north of New York City. The place was redolent of earlier centuries and its Dutch origins. We lived about a mile west of the Albany Post Road, which was traveled in the Revolutionary War days by George Washington and his Continental Army during the winter of 1777-78. The army had camped near an old mill two miles east of town. A plaque there commemorated that event.

Few of the homes in town dated from the 20th century. Some had the Mansard roof so popular in the 1870's. In one such magnificent pile lived Mr. Courtney, who sported a monocle and was always dressed in suit and tie. He looked the spit and image of the famous Eustace Tilley, the fop who annually graces the front page of the *New Yorker* magazine. Two blocks away lived Mr. Bogardus, the greengrocer. A small A&P store downtown was flanked by Mr. Bogardus' shop on one side and the butcher shop on the other. About four doors down was the Dutch Reformed Church. There was a rundown Sunoco gas station on the Post Road, but we always patronized the Texaco station that was part of Roy Ketcham's DeSoto-Plymouth salesroom and garage. During the war years and immediately afterward the showroom was always conspicuously empty. Mr. Ketcham sold whatever used cars appeared during the war, and customers snapped up any vehicles he could obtain in 1946 and '47 before they could be displayed.

My father car-pooled with other Texaco employees, driving the one car we owned, a gray 1941 Plymouth two-door sedan, one week out of four. We walked the mile to shop downtown, walked up the hill the 3 mile round trip to Mr. Van Voorhees' (note the Dutch names) farm to get fresh milk every week. I rode my bicycle to school downtown.

The car sat in the garage for days at a time when my dad wasn't driving to work, due to wartime gasoline rationing.

I was still too young to be fully aware of the circumstances, but in early 1943, my father again experienced the manic phase of his bipolar condition and went to a hospital in Poughkeepsie for treatment. The euphemism for what occurred was: "Your father hit his head on a radiator and he had to go to a hospital for treatment." Part of the therapy for my dad was to have an avocation, and he purchased a small fifteen foot sailboat. We sailed on it many times during summers in the mid and late war years. His doctors also urged him to take up pipe smoking, to help keep calm, but that was not good for his asthma. My parents set up a Victory Garden during the war years on a half-acre in our backyard. We enjoyed a wide array of vegetables, including fresh tomatoes and hybrid corn-on-the-cob, which my folks picked fresh just before eating. As a small kid I loved to eat at least two big ears of corn for dinner, as it was so delicious.

I have fond memories of the very upbeat approach that Frank Coleman had to life after his treatment in 1943. During spring and early summer of that year, he liked to play our big radio downstairs tuned to a volume loud enough that we could hear it upstairs in the bathroom, where Dad and I were getting ready for the day; he with shaving and me with washing my face and watching him shave, wondering when I would have to do such strange things. He played the morning music parade, so we heard lots of early Frank Sinatra, which wasn't our favorite, and such popular hits as "Mairzy Doats/" Mairzy doats and does eat doats and little lambs eat ivy, etc." Dad also liked to sing his favorite tune, perhaps from the twenties, namely: "Hey baby, don't mean maybe, hey baby, don't mean maybe now." This all sounded very playful yet adult, and was an unusual side to Dad's personality unveiled to me. Mama was busy fixing breakfast by then, so it was just a guys' sing-along morning.

I had several friends in school, including Henry Kohrhamer and Jimmie Chamblee. We rode our bicycles all over town, played pickup baseball games and generally enjoyed our mobile life-style. In mid-winter, we went sledding on hills near home and relished the freedom we had in those wartime years.

A small vignette gives an example of the sheltered life we lived in during the World War Two years. My parents seldom attended church, but they insisted that I go to Sunday school in the local Methodist church. The church had occasional social events, however. One time in late summer of 1944, the church board showed a newsreel account of events in the early days of the war. The film presented first the bucolic nature of life in Poland in the summer of 1939, and then proceeded to show the onslaught of the German army during their Blitzkrieg in September, gunning down many hundreds of soldiers and civilians alike, complete with the deadly Stuka airplane attacks, with sirens screaming. We watched no more than ten or fifteen minutes of this, and the audience, virtually as one, cried out, "Turn it off! This is too horrible." We all went home in a very upset mood, needless to say. I give this as an example of finding out what the war really meant to the combatants, not just something we had complained occasionally about with gasoline rationing, and meatless Fridays, etc.

Fishkill was like a southern town, writ small. We were predominantly Anglo-Saxon and Protestant, and there were no children in the Elementary School from minority backgrounds. Several African-American families, with few, if any, children, lived in a large unpainted old frame house, on the edge of a hill on the southern edge of town, leading down to the railway station on the New Haven Railroad branch-line from Hopewell Junction to Beacon, N.Y. The inhabitants were field laborers in large vegetable fields just west of our home on the west end of town. The field hands rotated their work time between Florida in winter and Fishkill in spring and summer. My parents were concerned that we keep out of harm's way and I was told to stay away from the fields. Oddly enough, my school friends and I had no interest in the fields or the workers.

The garbage man for the town of Fishkill was a black man, Frank Dolfinger. He drove an old beat-up Ford truck to pick up the trash every week. Frank was also the coal dealer. Our old home and many others in town were heated by coal, which Mr. Dolfinger delivered once or twice a year from his gleaming dark blue 1938 Ford truck. Being in the midst of a major war, and dimly aware of a coal miners' strike in 1943, I recall my parents talking anxiously about the effects of a prolonged coal shortage leaving us unheated in midwinter. Dad shut the furnace off in late evening, and the pipes on the second story of the house froze occasionally. Dad had to boil up a kettle of water and pour the water on the pipes, which meant much mopping up, but we got by all right. We also had a large fireplace downstairs in the living room, giving us plenty of heat in the evenings.

The contrast between our little town and Beacon and Poughkeepsie, 13 miles north of there, could hardly have been greater.

The four-track mainline of the mighty New York Central railroad ran up the east-side of the Hudson river from New York City, and on up, past Hyde Park, to Albany and points west on the famous "Water-level Route." Freight trains crawled along the "west-shore line," along the Hudson's west shore. In addition to the constant stream of freight and passenger trains, there was the river traffic: the Beacon-Newburgh ferry, sailboats and sightseeing boats of the Hudson River Day Line. My parents sailed their boat, a fifteen-foot long Maine Coast named the Puffin, out of the Chelsea Yacht Club, just north of Beacon. With such a large and tempting cool river to sail in, it was considered so polluted in the 1940's that we never swam in it.

We often encountered the big side-wheelers of the Day Line, crammed with sightseers in the war years, on boats with names evocative of the past: The *Alexander Hamilton*, the *Henry Hudson*, and occasionally, the Grande Dame of the line, the elderly *Robert Fulton*, with its big red walking-beam moving up and down, driving the large side-wheels that propelled her.

Perhaps this sounds idyllic, but it was far from it. The high humidity of the southern Hudson River Valley greatly aggravated my father's asthma, and in those days of no air-conditioning, we sweltered through long summers, cooled only infrequently by trips on the river in our sailboat, or in the aftermath of the occasionally ferocious thunderstorms that swept over us in the afternoons. The long and snowy winters were equally difficult to deal with.

Dad was fascinated by sailing, and subscribed to "The Rudder," one of the principal magazines in the sailing genre. I read some of the articles, and marveled at the advertisements for large powerboats, such as Chris-Craft. Dad, being a sailing purist, had only scorn for such frills as powered boats of any sort. He spent endless evening and weekend hours maintaining our boat, the *Puffin*. I helped paint and varnish the boat on occasion.

Life seemed simpler then. As we ate our breakfast soon after seven a.m., we looked out of our kitchen windows and occasionally saw Walter Overbaugh, one of the scientists at the Texas Company laboratory, walk by on Main Street the three miles to his laboratory in Glenham. This hike, while pleasant several seasons of the year, was most impressive when he continued to walk back and forth in the heavy snowstorms during the winter, bundled up against the cold.

4.
Midsummer trip to Montana in 1944

We had a pleasant interlude during the midsummer heat of 1944, with a memorable trip to visit my maternal relatives in Montana. As a young lad nearly six years old, I got a chance to travel across country from southern New York State on a nearly three-day each way vacation trip with my parents in July of 1944. The memories of the sights, sounds and smells of this trip have remained vivid in my memory, despite the passage of more than seventy years.

My parents saved up for this trip, as train travel was the only feasible way for our family to travel during war-time, the nearly 2500 miles from Beacon, New York to Havre, Montana, and my grandfather's home and nearby ranches. Grandpa Cowan helped out by mailing my parents a check for two hundred dollars (equivalent to about a thousand dollars now). My father, a research physicist for the Texas Company (later Texaco), was allowed two weeks' vacation, so my parents planned an itinerary that accommodated that interval. Early morning in mid-July, 1944, we departed from Beacon, boarding the Pacemaker, the New York Central's crack coach-only train to Chicago. Some family friends met us at LaSalle Street Station that evening and we went sightseeing the next day in downtown Chicago.

The long trip from Chicago to Montana began late the next evening with our 11:45 departure from Union Station on the first section (all-Pullman) of the Great Northern's Empire Builder from Union Station. I climbed into an upper berth and slept the best sleep of my life, being gently rocked by the big heavyweight Pullman car on the Chicago, Burlington and Quincy main line to St. Paul, Minnesota. Mid-morning the next day in St. Paul, one of the high points of the trip occurred when my Dad took me to the front of the train and we watched the crew changing locomotives prior to the departure of the Empire Builder for points west. Walking alongside the big sleek Vanderbilt (round tank) tender, the red and white shield showing Rocky, the goat, invited one and all to visit Glacier National Park. Then we saw the shining black Great Northern Railway 4-8-4 locomotive. I later found out this was one of the famous S2 Northerns built by Baldwin in 1930. The engineer looked down at us and asked if I would like to take a look around in the cab. Would I! My dad hoisted me up the steps and I had a look at the gauges, valves, and was awed by the sounds of pumps, the fire in the firebox, and the smells of hot oil, steam and the feeling of pent-up power waiting to make the departure. The engineer gave me a large handful of cotton waste (for cleaning the gauges) that I proudly took back to our compartment with me.

The trip to Havre, Montana took a full day and two nights. Visits to the diner were fascinating, with hot food served on china and spotless white linen tablecloths. Walking to the diner, we went through several cars, some of which were like visiting a turn-of-the-20th century home. One Pullman car in particular had mahogany paneling and lots of polished woodwork with frilly lacework curtains on the windows. The ornate lamps swaying overhead were fascinating as well. My father turned to us and said: "Isn't this fantastic?" Years later, I still wonder if this car was from the old Oriental Limited fleet from the early nineteen twenties, pressed into service for the peak passenger loads during war-time.

Change was in the air. On many side tracks we saw big Omaha Orange and green diesels, pulling long freights that were waiting for us and the second (coach) section of the Empire Builder to pass. I learned later that the Great Northern had a large fleet of 5400 horsepower FT (Freight) locomotives (made by the Electro-Motive Division of General Motors), second in number only to those of the Santa Fe Railway. Diesels were a real novelty to me, because the trains of the New York Central were still all steam, as was true for the western part of the New Haven Railroad, which crossed the Hudson on the high bridge at Poughkeepsie, a few miles north of Beacon.

My grandfather, William T. Cowan, met us at the Havre train station in early morning and took us home in his large dark green four-door 1941 Packard to Box Elder, a little town of ca. 200 people twenty-two miles southwest of Havre. We were welcomed by Grandma Cowan, who called me "precious." In return, captivated by this warm, loving and small (five feet tall) lady, I called her Preshie, a name that I used for the rest of her life. Shortly after being greeted by Preshie, I met her younger sister, Aunt Effie, who was even smaller, so I called her "Another Preshie."

I also was captivated by visiting Aunt Minnie Thompson (Grandpa's older sister) and Uncle Albert ("Op"), her husband, who lived in the Cowan ancestral home about a five minutes' walk from the General Store. Uncle Op entertained us by playing games and giving us "horse rides" on his knee.

Time flew by rapidly as we visited Grandpa's General Store, and cattle and wheat ranches. We rode horses, and did some trout fishing in the Bear Paw Mountains twenty miles south. Mid-July was peak time for the wheat harvest coming in to the Cowan and Son elevator in Box Elder. It was fun to watch the local boys "coopering" the boxcars (nailing up rough-sawn pine doors that covered the open doors of the cars to enable the transport of several thousand bushels of wheat per car).

During the war, as was true afterward, the boxcar shortage in the plains states during the wheat rush from July through September was a source of concern to the ranchers. Grandpa Cowan could ship as many as a dozen or more cars per week, but seldom received more than eight. The cars were picked up by a way freight that came through Box Elder in mid-morning on its way to Great Falls. The cars of wheat continued on their way via the "high line" across the top of Montana to their ultimate destination, the Centennial Mills in Spokane, Washington. Attending church on Sunday morning was a ritual in the family, and we had a big dinner afterward. Grandpa read a lesson for the day in "The Upper Room," a Methodist religious journal, followed by grace, followed by the meal. Silence was expected during that meal, however, as Grandpa Cowan listened closely to the prices of wheat being broadcast in early afternoon on his big Philco radio.

Several trains went on the line from Havre through Box Elder to and from Great Falls every day. In addition to the freights, pulled by 2-8-2 locos, called "Mikados" by railroaders; there was the "Helena train," a short local passenger powered by a shiny dark green Pacific (4-6-2 wheel arrangement) locomotive with a baggage/RPO car, a coach and a parlor car. It took all day to travel the 200 miles to Helena. The Helena passenger arrived at Box Elder at approximately 8 a.m., and was followed by the westbound morning freight about one hour later. Early in the morning, at about daybreak, 4:30 a.m., the "Galloping Goose," a fire-breathing gas-electric spouting blue and orange flames out its exhaust stacks, came through town, carrying express mail and passengers to arrive in Great Falls by mid-morning. My father took that train at the end of our visit to Great Falls and then flew down to Denver and on to Port Arthur, Texas, on business for Texaco. He met us back in Beacon after we returned home the way we had come.

The trip made an indelible impression on me for several reasons. It was a chance to get to know my grandparents and several aunts, uncles and numerous cousins. The ranching and wheat business was a totally new experience. However the most intriguing aspect of the trip for this budding rail-fan was the incredible diversity of motive power, rolling stock and activity in the railroad industry late in World War Two.

Once the war was over, my parents never again traveled by train. In August 1947, the family moved to southern California, and made many vacation trips from San Diego to northern Montana, but always by car. I have since traveled by train, once in August 1954 on five different railroads on a trip home to San Diego from Box Elder. In later years, I have traveled occasionally on Amtrak, including two trips on Amtrak's Empire Builder in the last twenty years, but never felt the sense of wonder of that first trip.

Another vacation trip that we took was by car in August 1945. My parents had saved their gasoline ration coupons assiduously for years, and then found them useless once VJ (Victory over Japan) Day was declared in early August. We rode in our Plymouth two-door sedan up the East coast to Maine, and arrived at the Casco Bay area just north of Portland. We took a boat over to Birch Island, and stayed in a summer cottage that was owned by family friends from my Dad's days at Bennington College, Harold and Lenore Gray. It was mostly cloudy and cool. I recall being scandalized at my young age that mom and dad would go skinny-dipping in the Atlantic from a rocky beach near the cabin. It was so cold that I declined the offer to join them. We took frequent walks across the island through clouds of voracious mosquitoes to a small store to buy groceries. My mother was always one to keep active, even on holiday. She managed to wash and wipe dry the hundreds of windowpanes on the well-illuminated cottage.

After our one-week sojourn on Birch Island, we returned home by an inland route, visiting my dad's friend, Dr. W. A. ("Andy") Saunders, a noted physicist at Amherst College. We also visited William Trufant Foster, former first president of Reed, at his home in East Jaffrey, New Hampshire, just on the Massachusetts border. Dad called him "Uncle Bill," and happily recalled the times when he and his parents visited the Foster family in the President's House, on the southwestern edge of the Reed College campus, when my father was a young boy. Foster kept a small herd of goats grazing his large lawn in what was then very open country, with few homes to be seen for miles. Mr. Foster seemed somewhat subdued, but had made a success as an investment banker on Wall Street.

From an early date, my parents saw to my musical education and experience. They took me to the new musical, "Oklahoma!" by Rodgers and Hammerstein. In the spring of 1943 we drove down to New York City for the show. The big cast and robust nature of the play with its numerous songs awed me. In 1944, my mother got tickets for a Saturday afternoon matinee of "Hansel and Gretel," performed by the Metropolitan Opera. That was a bit scary for a nearly six-year-old, but such is the nature of Grimm's fairy tales. Another time, my mother took me to New York City to see the famous Ringling Brothers circus, performing in Madison Square Garden. I especially enjoyed Emmett Kelly, the sad faced clown, who walked onstage with his big black umbrella, squirting water up and over him, as he walked blithely around.

The following episode is fraught with Freudian overtones, but is still clear in my memory. In early January 1947, I had gone to bed soon after 8 p.m. and was sound asleep as usual. In the middle of the night, I heard loud conversation and my mother giggling in their bedroom. Being naïve as always, I walked down the hallway and tried to open the door into their bedroom. My father called out in his loud baritone voice, "David, go back to bed right now." I did as ordered. At breakfast the next morning, my parents were very subdued. Lo and behold, my sister Margery was born nine months later. The timing was right; I never mentioned it to my parents. The episode stays in my mind for what it is worth.

5.
Big change, move to San Diego, California

*M*y father, a Ph.D. physicist ever in pursuit of the main chance, was desirous of relocating to the west coast. After interviewing early in 1947 to be director of the Naval Electronics Laboratory (NEL), beginning in late August, on Point Loma, in far-off San Diego, California. I recall my parents' many anxious conversations about this big move: would it be successful, and could we even afford it? Decades later, I realized we were living a version of one of Wallace Stegner's novels, such as *Angle of Repose*, in which the main protagonist was always pursuing a dream, trying to succeed against considerable odds against him. I wondered about my mother's anxiety. With my dad's previous history, she must have worried that this big change of scene and increased responsibility was unsuitable for him. A year or so later, her worst fears were realized.

To provide funding for the move, there were a few war bonds to be cashed in, and little else. Rent on our big house had cost $50/month, plus utilities, during the World War Two years. The rent had increased to $60 in 1946, which my irate father noted to our landlord was a 20 percent increase and obviously unfair. Armed with these financial stimuli, my dad was determined to move us in the ultimate in do-it-yourself style: buy a truck, load it with all our furniture, and drive it out to San Diego. My mother, seven months pregnant with my sister-to-be Margery, would fly out to visit her parents in Montana en route, and then join us once we had found accommodations in San Diego.

One positive note remains from my last few weeks in Fishkill, relating to peer pressure or approbation. Several of my friends were concerned about my move clear across the country; would I ever see them again? I never did see any of them again later in life. A small group of us rode our bicycles out along the road to

Wappinger's Falls and dropped in on Gary Mombello, a somewhat precocious son of one of the few MDs in town. Being late summer, we were looking ahead to the coming school year. Gary asked me, "David, what are your plans?" I casually mentioned that my family and I were soon to move out to the west coast, to San Diego, California. He beamed, and said in his mellow tenor voice, "Wow, that is great!" That was the first time any of my peers had congratulated me on making such a big move.

Before the trip to California, however, my parents resolved to see the mid-Atlantic region of the United States, using up my father's two weeks of vacation time before we moved west. In early July 1947, we drove down to the New York City area, and went on the Pulaski Skyway, a long bridge over suburban homes in the Jersey City area of New Jersey. Unfortunately, we were in a "fender bender" on the rain-slicked cobblestone surface of the bridge, being the third car in a three-car pileup. After a few repairs, we went along the New Jersey shore, then to Delaware and, via ferryboats, to coastal Virginia.

We saw the islands of Chincoteague with its wild horses, and Assateague. Walking across tidal flats at low tide, we walked on thousands of small crabs, which were teeming on the surface of the mud. We met a crab fisherman, who commented that the crab fishing was very good that year. Being young and impressionable, I said I thought we probably saw more than one hundred thousand crabs that day. The man replied: "There are millions of them out here." He noted that ten hundred thousands make one million. Dad immediately affirmed that fact.

The most fun part of the trip was saved for last. We drove inland to Washington, D.C. and found a hotel to stay in (The Mayflower), and drove around to see the famous sights: the Lincoln and Jefferson Memorials, climbed up the 555 steps of the Washington Monument, and also visited the White House. It was hot and humid, and we were wilting some of the time, but it was no worse than the mid-summer climate of Fishkill. My mother had written ahead to a distant relative on her side of the family, Mr. Robin Cox, who was a Representative from Georgia. We sat a few minutes in the visitors' gallery of the House of Representatives, and saw very little action on the floor, which was as true then as it is now. That was a lot to absorb for a young kid who had just finished the third grade. Exhausted, we headed back home to New York State.

6.
The epochal trip to California:
The summer voyage of "Big Sister"

The idea seemed good, but where could one find a truck in those pre-U-Haul days? Mr. Ketcham had sold a farmer a 1.5-ton Dodge stake (wooden sides) truck in 1946, which was all he could get from Detroit but not the pickup truck the man had really ordered. My dad traded in the Plymouth sedan (complete with its badly dented front grille), plus the proceeds of a Victory Bond on the truck. Ketcham found a new pickup for the farmer, leaving him with a used car to sell: the ultimate win-win situation.

Dad built up the sides of the wooden stakes to eight feet in height, hired a packer from a moving company who loaded our living room grand piano, my mother's pride and joy, up next to the cab, my Lionel O gauge train layout, and all the household furniture totaling three tons, or double the rated load. After covering the load with big tarps and much nylon rope to cinch it down, we were ready, we thought, for prime time. This adventure, the total uprooting of a young boy and his family, was one of the epic events of my life, one that I can recall to this day on an almost hourly basis.

There were some positives: the truck, being only one year old, had low mileage, and was in prime condition. My dad had driven large farm trucks in Oregon in the past, and was familiar with the need to double-clutch between first and second gears of the four-speed gearbox. The minuses were: the weather (we chased a heat wave across country in mid-August 1947, cooled only by two open windows and the large air scoops at our feet, which one could open wide to draw in outside air). With the overload, our six-ply tires were hard pressed, as we had three blowouts en route.

I climbed into the cab of the big blue Dodge ("Big Sister," named hopefully for my soon- to- arrive sibling, little sister Margery) with my faithful Border Collie Tippy. [insert picture here] Dad cranked up the big six-cylinder engine, and we set off on August first. In those pre-interstate highway days, our Texaco map routed us around most urban areas, following the byways made famous in the 1970's as *Blue Highways*, by William Least Heat-Moon. We crossed the lower part of western New York and traveled on part of the Pennsylvania turnpike, turning off at Carlisle to head across western Pennsylvania. Our first night was spent in a noisy motel room that seemed straight out of *Lolita*. Ironically, Nabokov set that novel in the late 1940's. We heard the diesel tractors gearing up and down on the hills, and I slept very little. We awoke early and continued on the winding road, just missing a truck wreck by less than a minute. We could see the overturned tractor rig with its wheels still turning after it had landed on its cab. Our second day took us halfway across Ohio to Columbus where family friends put us up for the night. Both dad and I learned the hard way that it is better not to eat much breakfast when setting out on a very warm twelve hour day of driving on crowded two lane roads behind equally heavily-laden trucks for much of the time. I was fascinated by watching the chains dragging under gasoline tanker trucks, the chain bouncing along leaving a trail of sparks that was supposed to prevent static buildup and the cargo's explosion.

    The next several days dragged by as we strove to travel up to forty miles per hour, and at least 400 miles per day as we crossed the flat lands of Indiana, Illinois, across the Mississippi and on into Missouri and Kansas. Rest breaks were needed for Dad every two hours and stops for gasoline at least three times a day (what fuel mileage did the truck get? I never asked and Dad seemed to prefer not to know). We had to make many detours around road construction projects.

    On one particularly rough stretch in eastern Kansas, we pounded over a very rutted road, and knocked the air filter loose. Clouds of black smoke billowed out from under the hood. After retrieving the air filter, Dad poured more oil into the oil bath and we were on our way again. Soon after that, we went for a swim in a gravel quarry, and drank ice-cold Grapette sodas. That was close to heaven in the Kansas heat. We spent that night in a hotel in Marysville, along Route 36, which also was along one of the Union Pacific main lines, headed to Denver. My dad had trouble sleeping, but those big steam freight and passenger engines put me right to sleep.

Our next day was a make or break run across over 400 miles of Kansas on route U.S. 36 and then on directly into northern Colorado. This was an occasion for celebration because we had a rest and recuperation half-day with my grandmother's sister, "Aunt Effie," in her apartment on Capitol Hill in Denver. Her home-cooked soups and some shade and relative coolness revived both of us for the final run on to California.

We headed south to Santa Fe and Albuquerque, then turned west on fabled Route 66 across the high plains of New Mexico and forested lands of northern Arizona. The purple haze on the majestic mountains of New Mexico was really exotic. Did we detour to see the Grand Canyon or other scenic delights? Not on your life! This was a battle to the finish, and apart from camping out on our air mattresses some nights by the side of a byway, we were ready to make the crossing into the Promised Land. Such was not to be so easily attained. We sped across the long downgrades in western Arizona, past "The Needles," into the Mojave Desert of California. Local thunderstorms were visible from over fifty miles away in the late afternoon. To enter the Golden State, we first had to pass through the Agricultural Inspection station at Needles. People in cars were being waved through, after declaring that they had neither fruits nor vegetables with them. Upon making a similar declaration, my dad was directed to pull over into the far side of the parking area, in a spot where he could open up the entire load. This was a several hour process, and I was no help at all, asking why this was happening, and being told to "look after the dog."

After four hours in the desert heat, the inspection station official checked it all over, and then the load had to be recovered and cinched down. On top of all that, one of the dual rear tires had blown just outside of town (our third blowout of the trip; replacement cost was $75, or the equivalent of nearly $400 now). We finally bedded down in a state campground after midnight and arose early the next day. That was to be one of the epic runs of our trip, traveling across the desert to Barstow and Victorville, then over Cajon Pass, and down into the Los Angeles basin. Fortunately, the trip that day went fairly well. We cruised downhill on the west side in third gear, making over 400 miles that day as we sped through the citrus groves west of San Bernardino and south of Los Angeles.

We then drove straight West to Highway 101 and had a swim in the dark blue Pacific Ocean south of Long Beach. How refreshing that was! Unfortunately that day was Tippy's introduction to the Wild West. He had never encountered prickly pear before and got a nose full of cactus spines. We found a veterinarian (on a Sunday!) who pulled several dozen out, and set off on the final hundred miles to San Diego, more exhausted than joyful but at our destination at long last. The trip took 12 days, three new tires and very little travel money remaining in Dad's wallet.

7.
The early years in San Diego

*T*he first few weeks in San Diego passed at a relatively slow pace. Finding any family housing even with Dad's defense-related job, was next to impossible. We stayed in the Hotel Ocean Village; small rooms perched on a pier at the North end of Ocean Beach. One evening, we went out to eat at a restaurant on West Point Loma Boulevard. We had hamburger steak, mashed potatoes and a stalk of celery, for $1.05 each. Dad protested that such a high price was highway robbery. Quite a contrast to prices nowadays!

After a two-week search, Dad found us a temporary home in a duplex in a housing project built during the war, fifteen miles inland in Linda Vista. My mother joined us, giving birth to my little sister in Mercy Hospital on October seventh. Our life began anew in early November in Ocean Beach on Point Loma, 4544 Granger Street, only three blocks from the ocean. My parents lived in that house for over forty years.

In hindsight, no other cross-country trip was ever so long and so memorable. A short afterword: The market for used trucks in San Diego was very depressed in late 1947-1948. My grandfather in Box Elder, Montana told my dad to hold out and not sell the truck for the fire-sale price of $1200, when he could get nearly twice that in Montana, as large wheat-hauling trucks were scarce. My father parked the Dodge for a full year (on a vacant lot next door) and finally drove it up to Montana with the family and our old 1934 Chevy two-door sedan riding in its bed (for the return trip) in August of 1948. We went up the San Joaquin Valley to Sacramento and then over the high Sierra Mountains following the Southern Pacific (SP) railroad, on US Rte. 50. In Dutch Flat we visited an old friend of the family's, Bill Adams, who was a salty old character, making his fame writing novels such as *Ships and Women*, which was popular back in the 1930's. I vividly remember the sound of the big articulated cab-forward Mallet (steam) locomotives on the SP that hauled long freights up and over the Sierras to Reno and farther eastward.

Upon our arrival in Box Elder, Montana, Dan Bitz, one of the wealthy wheat farmers in the area, made an offer and purchased the truck for cash. All the wheat ranchers in the area were confirmed Ford, Chevrolet or International owners. Mr. Bitz was so impressed with the toughness and hauling capability of our Dodge that he assembled a fleet of half a dozen big Dodge wheat-hauling trucks over the next ten to fifteen years. At last report, in 2010, he still had the 1946 Dodge parked out back of his ranch house in Hill County, no longer in use but in a place of honor.

Another aspect of life at Box Elder was the big Cowan and Son general store, which was a true relic of the 19th Century (established in 1888). Every county in the west had one or more general stores, which stocked about everything except fresh produce for the farmers and ranchers within a twenty-mile radius. They sold on credit, and expected to be paid after harvest time. In later summers, I used to help stock the shelves and do general sweeping up if I was not needed elsewhere during several visits to my grandparents. Some aspects of the business seemed frozen in time, perhaps from the 1920's. Grandpa Cowan sold Silver gas, the local brand from a refinery in Great Falls. In early 1948, Phillips petroleum bought that refinery out, so the gas pumps had the bright Phillips 66 symbols. There were three pumps, one on each end with regular gasoline, powered by electricity, and the middle one, with Ethyl, had a tall glass column, that was pumped by hand to a maximum capacity of ten gallons. The gasoline was then dispensed by gravity into the car's gas tank.

During our visit in August 1948, we also had a cross-cultural experience. The entire extended family rode up to Rocky Boy's Reservation on the western slope of Mt. Centennial, south of town. The reservation, comprised of Cree Indians originally from

Canada, had several hundred inhabitants, and they were served by Cowan and Son's store on the reservation. Grandpa was welcomed as an old friend, as he and his father David Cowan were known by the Cree as "Little Bones" and "Big Bones," respectively, from the early days in the late 1880's when they gathered and shipped out many boxcars of buffalo bones from herds that had been slaughtered on the grasslands in the 1870's. The bone sales to fertilizer factories in East Saint Louis, Illinois subsidized the Cowan and Son firm in the first few years until they had built up a herd of beef cattle. During our visit to Rocky Boy, we watched a Sun Dance, with many singers and dancers. It was all very exotic and hard to comprehend for a young city kid.

Another aspect of being in a farm-oriented economy: Cousin Bill Cowan had gotten his driver's license at age eleven, to help operate the company's cars and trucks. In 1954, when I was 15 and not yet licensed to drive in California, Uncle Bill considered me old enough to drive his old 1946 Chevrolet gasoline truck to help in making deliveries to ranchers. The Chevy had two gasoline tanks mounted on the flat bed, each with a capacity of 293 gallons. Driving on back roads up to the western side of Centennial Mountain to John Sheehy's ranch with one tank each of regular and ethyl gasoline was a memorable experience. George Colligan, a roguish old Irishman who worked for my uncle, kept me company. There was a lot of play in the steering mechanism, and one had to try to anticipate what the front wheels would do in the often-rutted gravel roads.

8.
Major setback for Frank Coleman

Soon after our return from Montana, in the fall of 1948, Francis's bipolar condition reactivated into a manic phase, as evidenced by his spending hours listening to signals on the shortwave-frequencies of our big radio at home, convinced that some secret messages were being sent to him. At that point, it was quite a wrench to send him off for treatment, as we knew of no local sanitaria. Dad went to a large state-run facility, Patton State Hospital, near Redlands, California. This incarceration lasted for over ten months, again replete with electric shock therapy. My dad's retired parents, Norman and Ethel, came down by car from their home in suburban Portland, Oregon, and provided some financial assistance and help in caring for my year-old sister Margery. I rode my bike to elementary school, and managed to learn how to get along with kids in a big suburban school.

Every other Saturday we took long three-hour trips (each way) from San Diego to visit Dad in the hospital, crammed into my grandparents' small two-door 1941 Chevrolet coupe. The adults met with the doctors, while Margery and I waited outside under the pepper trees on the spacious grounds of the facility. There were lots of unknowns for all of us. Dad watched us reproachfully from afar. As far as he was concerned, there was nothing wrong with him, and he was doing it mainly to please my mother.

In order to have some income for our family, my mother got a teaching job at a private elementary school just three blocks away from home. Her monthly salary was $100, before taxes. This was not enough to keep us going, so my grandparents provided some financial support and minded my sister Margery part-time. Grandmother got a job teaching English at the Francis Parker School across town. My mother applied to teach English courses in the

San Diego County area, and got an opening at Grossmont High School, out past La Mesa in the eastern part of the county. She carpooled with others and managed to boost her income significantly. She also took courses in English literature and got her Master's Degree from San Diego State College in 1952. Her Master's thesis, entitled: "Scattered Radiance," was an evaluation of the writings of Henry James.

Meanwhile, Alice turned up an opening for an English teacher at La Jolla High School, which was half the commuting distance of her previous job. By 1955, she moved to the new Mission Bay High School, and stayed there until the end of her teaching career in 1979.

The unsettled situation in our family seemed atypical compared to many of my friends' homes in Junior High and High School. We know now that the prevalence of mental illness in the USA is about ten percent of the total population, so it is far from unusual. I occasionally asked my mother about her impressions of my father's bipolar episodes, and the background came out, piecemeal, over the years. To put it succinctly, Francis Fleming Coleman was an overachiever and near-genius-IQ student, who was home-schooled by his mother, Ethel, through the eighth grade. This was after grandmother had a series of miscarriages after giving birth to Frank in June of 1908. Thus she concentrated her love and attention on Frank, who managed to get straight As all through high school after entering it about two years ahead of his usual age cohort.

To compound the situation, Ethel was a single mother for some of those years, as Norman, my grandfather, was often away on travel. For example, he served as secretary to the YMCA in Paris during World War One. In the early 1920's, my grandparents adopted two children, Lewis and Barbara, which was quite a venturesome and trailblazing thing to do back then. Dad was then encouraged to take a year off (in 1924) and work in the American Can Company plant in downtown Portland, working with a rough and ready group of laborers. One fellow, a foreman there, delighted in throwing a knife at Frank, hitting the wall nearby him as he walked along with boxes, etc., just to test his nerve. This never injured dad, but it was telling enough that he recalled this among a host of other experiences designed to "toughen him up" at the time of his first bipolar episode, or "nervous breakdown" as it was termed in the 1930's and 40's.

Mama mentioned him shrugging off a large dose of narcotic injected into him at the sanitarium in Hartford, CT, in late 1938, as he tried to work through the bewildering array of conflicting memories that flooded through his mind. One of them was rather chilling to me: as his train left Portland on his trip to University of California, Berkeley, his mother ran alongside the train, and yelled out to him at his open car window: "Remember your I.Q.!" This was obviously a young man from whom much was expected, and he was kept well aware of it.

The frustrating, but understandable, situation for my sister and myself was by the time we were fully cognizant of some of the problems Frank had to deal with through life, he was usually very quiet and subdued, except at mealtimes, when very proper table etiquette had to be observed, and little was discussed except for current events in the news, etc. Before Frank was put on a lifetime prescription of the drug Melloril in 1958, Dad would occasionally sing small jingles from radio advertisements, as a bit of humorous byplay. He had a propensity for coining puns, and had a great penchant for coming up with unusually difficult city and town names when we played that game on long car drives. As those who have lived with family members with bipolar disorder, the big concern was: what sort of mood will he be in today?

A further insight into the situation between my mother and father developed late in my mother's life, when Margery and I had a chance to visit in the 1980's with Eckoe Ahern, a life-long friend of Alice's from the days when they both taught English at Grossmont High School, at the east end of San Diego county. One day in

1949, during Frank's long hospitalization for his bipolar disorder, Alice and Eckoe sat down to chat over lunch in an outer courtyard. Mama said, bursting into tears: "What happened to the Greek God I married?" She had very high hopes for her life with Frank from the time they were married in 1934.

One major feature of my life growing up contributed to our family's mental health. My mother Alice was a very demonstrative and loving person. She made sure that Margery and I got lots of hugs and encouragement, along with the usual expected tasks of daily chores and getting homework done, to keep peace in the family. Perhaps because of my dad's overly pressurized early childhood, my IQ was never brought up, and although grades were expected to be good, I got mostly A grades in high school, particularly during the two years when I played in the band. My GPA was a bit lower in the usual college prep courses in high school. In my years at Reed, my overall grade record averaged a C-minus. The Biology Department advisors sent warning letters to my parents but never said a word to me, except for the occasional "bring the grades up," exhortation by Dr. Lewis Kleinholz, adviser in my first two years at Reed. What a strange system! It was the unmentioned elephant in the room, as it were, as Reed proudly adhered to the bizarre system of issuing A-plus, A, and A-minus grades on term papers and exam papers (where A meant average). The idea was that we should not be concerned with grade levels, even though most of us were headed to graduate school after the grueling four years at Reed.

My father returned home in late summer of 1949 and gradually began to work part-time on physics projects at the Scripps Institution of Oceanography, located along the coast north of La Jolla, in isolated splendor decades before the formation of the University of California, San Diego campus. In May 1950, my grandparents returned to Portland by bus, generously giving their 1941 Chevrolet car to my parents.

9.
The milieu in grade school in the late 1940's in San Diego

The San Diego school system was very generous to its grade-schoolers. We went on a five-day trip to a camp in the Cuyamaca Mountains east of the city, for a nominal fee. Another activity in the spring semester of the 6th grade, was an afternoon field trip to experience a plane ride over the city. We flew in a Convair 240, a two-engine propeller plane up over Point Loma and back over the vast expanse of San Diego Bay. We took a city bus to Lindbergh Field and back and felt very adult in our final weeks of grade school. We had a discussion in class about what we saw, and Miss Jones, our teacher, asked what we saw as we flew over the ocean west of Point Loma. Students suggested boats, waves, etc., and she said she wanted a particular object. I suggested that the sea surface looked like rough glass. Miss Jones delightedly said that was exactly right.

There were breaks occasionally in the family tension during our early years in California. In the evenings in those pre-television years (we did not get a TV, a small 12 inch portable Philco, until my senior year in high school, 1955), my father would read aloud from such family favorites as *"The Wind in the Willows," "Dr. Doolittle,"* and *"The Irish R.M.* (Resident Magistrate)" in his warm baritone voice.

My parents joined a Madrigal group, singing songs from the Renaissance era, with a number of friends from San Diego State College. I was considered too young to baby-sit in the early 50s, so my folks took Margery and me along; she slept in a back room and I watched television with other children of the singers. The group played recorder music; it was fun to hear the sounds of early Renaissance music.

10.
Musical influences in the Coleman household

The strong musical influence in my family was driven largely by my mother, who loved to be active in various projects. During my high school years, she sang in a choir with the San Diego Symphony, which played such pieces as *"King David,"* by Arthur Honegger. Another year, she sang in a group that performed some J.S. Bach Oratorios. The noted Bach historian and musician, Julius Herford, came out to San Diego from New York City and played Bach's *Well-tempered Klavier*. My mother invited Julius and his son Peter, who was in college at the time, out to our house for a meal on her fine china, and the discussion about all things Bach waxed eloquent. Peter went on to have a stellar career as an organist and keyboard artist in the New York City area.

Another aspect of cultural life in the San Diego area was always in the background, because it was omnipresent in our lives. We had season tickets to the Shakespeare plays that were presented during the summer in the Old Globe Theater in Balboa Park in the hills just above downtown San Diego. We saw many productions of *"As you like it,"* and *"Love's Labor's Lost,"* and the occasional production of *"A Midsummer Night's Dream."* The very tall and muscular young actor Jon Voight played the sprite, Ariel, one summer, impressing us all. There were songs and dancing on the green by young lads and lasses, accompanied by recorders. Imagine the impact of this on young children who scarcely knew any other music than what was on the top forty on the music hit parade on the radio. Certainly my sister Margery and I always had a chance to sample the offerings in the classical music and drama performances during the summers and also at chamber and orchestral music concerts during the school year.

In the summer of 1953, my parents took a break from traveling back home to Montana and had a quiet summer respite. To keep me occupied, my mother urged me to attend a music camp. I attended the Idyllwild music camp, run by the University of Southern California in the mountains above San Jacinto, east of Los Angeles. I lived in a large wooden dormitory, participating in various rehearsals of the Idyllwild Symphony, and also playing bassoon in a woodwind quintet. We played a little piece by Debussy, entitled "*Le petit negre.*" This was helpfully translated as the "little nigar." This was one year before the Supreme Court ruling on Brown vs. Board of Education. The notion of persons of African-American descent seemed very much in the background in that era. The conductor of the symphony was Noel Leon Arnaud, a composer of music for films in Hollywood. We performed Bizet's L'Arlesienne Suite, and Brahms' First Symphony. We also had a surprise visit from Meredith Willson, composer of "76 Trombones," who played the flute part in one of our orchestral pieces. In all, it was a fascinating view of how the other half (children of rather well-heeled teenagers in California) lived.

In my public school years, K-12, few teachers were memorable. In the eighth grade, my English teacher, Eugene Storm, was quite a standout. He liked to declaim poetry in his booming baritone voice. Edward Noyes' "The Highwayman" was a favorite of his, as was another by an English author, about the Etruscan chieftain, Lars Porsena. The stanzas went something like this:

> Lars Porsena of Clusium,
> By the nine Gods he swore,
> That the great house of Tarquin
> Should suffer wrong no more.
> By the nine Gods he swore it,
> And named a trysting day,
> And bade the messengers to ride
> North, South, East and West,
> To summon the array.

Mr. Storm was also wont to ask the students what they knew of Church holidays, and other events seemingly far removed from grammar and the usual subjects of English. One day he asked: "What is the most sacred ritual in the Christian church?" This was met with complete silence, then one or two students volunteered Christmas and Easter. He said, no, that was not the answer. He said it was the Feast of the Circumcision. At least half of us were bewildered or didn't know what he was talking about. The other half seemed embarrassed. I was in the former category and went home to look it up. Needless to say, he had few disciplinary problems, because we always wondered what he would come up with next.

During class one day, Mr. Storm asked us how far apart our homes were from ones nearby. Person after person said: "about one hundred feet." One guy, John Tate, said: "That's my situation now, but a couple of years ago, we were fifty miles from the nearest neighbor." Storm seized on that, and said that's what was typical for this country just one century ago. Tate explained that he and his family lived in the Texas Panhandle, along the Canadian River, and that sparse population was all the arid grassland would support. What an interesting aside to have in the midst of a Junior High English class.

In the midst of my restless and uncertain teenage years, Dean McBride, the pastor of our church, the Point Loma Presbyterian church, was an inspiration to me. I rode my bicycle the two miles to the church late one afternoon, after finishing my paper route, and visited with Dr. McBride. He was the epitome of a caring minister, with the bottom line: God cares about you, and what you do is important, and don't forget that! There was no willing one's way to wealth and fortune, which seems to typify so many churches now. His message still resonates with me, which is reassuring.

Douglas Hanna, the teacher of our high school group at the Point Loma Community Church during my sophomore to senior years, was very influential. Several times a year, he would say in his wonderful Belfast Irish brogue, "Boys and girls, young men and women, you have your whole lives before you. What a chance you have to develop a career and be a success in life!" This message was indeed an inspiration in our rushed lives, as we were changing both physically and educationally.

Mama was a life-long correspondent, sending out an extensive weekly letter to the two sets of grandparents in Montana and Oregon, and to the two great-aunts, Bertha and Louie, in Victoria, B.C., Canada. This enabled her to keep recent memories fresh, and to keep in touch with what may have seemed like a more normal existence elsewhere. Margery and I were expected to write letters at least weekly, when we were away for summer visits when we were young, or at college later on. Long after Fran and I were married, a weekly letter was expected from us. This is not so surprising in retrospect, because telephoning in those long-ago days of the Bell nationwide system was costly, and reserved for times of major crises.

My parents were ardent internationalists, and showed their interests in world affairs by joining an International House group in the San Diego area. During my father's years pursuing a Master's degree in Physics at UC Berkeley between 1929 and 1931, he stayed in a dormitory, International House, which housed several dozen students from a wide range of foreign countries. He contacted a few friends in the San Diego area and took the lead in hosting monthly meetings, several of them at our home in Ocean Beach, during the early 1950's, when I was in High School. It was enjoyable to meet so many people from Europe, Asia and the Middle East, several of them visiting in San Diego during sojourns on one or more of the naval bases in town. A young Egyptian fellow, Sayed El-Wardani, mentioned his great distress at the treatment of Palestinian refugees after the formation of the country of Israel. I knew several Jewish students at Point Loma High School, and Sayed's account of the vicissitudes of the refugees was in marked contrast to their comments about the new country of Israel.

After a few years, Dad was elected president of the International

House group. My parents put a lot of energy into entertaining the large group, which at times approached twenty attendees per meeting. When Dad had another bipolar episode in 1958, my parents dropped out of the group and became more limited in their social interactions.

During 1958-59, my mother co-edited an anthology, entitled *Introducing Poetry*, with her close friend and former Master's degree advisor at San Diego State College, Dr. John Theobald. Working on this book entailed voluminous correspondence with poets around the world. I recall her being concerned when I mentioned that Ezra Pound was rumored to be insane and out of touch with reality during his many years of incarceration in St. Elizabeth's Hospital in Washington, D.C. She noted that she had an extensive correspondence with Pound in the process of editing the book, and found his letters to her to be very lucid and insightful. The poetry book had modest sales in the US. However, many thousands of copies per year were sold in Canada, where some large high school and college systems adopted the book as a text.

Contemporaneously with this effort, my mother served as a reader of English essays for the College Board examinations, with the Educational Testing Service, in Princeton, New Jersey. All of this activity resulted in a significant supplement to her annual salary from the San Diego Unified School district, which was never greater than $22,000, even with her Master's Degree and considerable seniority. Once she retired in 1979, she became active as a docent with the San Diego Museum of Art, conducting tours once or twice a week in the Timken Gallery of that museum.

It was apparent to my sister and me that Mama could have been a college or university professor if her situation in life had been different, or she had grown up in a later generation, under less demanding circumstances financially. I was amazed that Alice's older sister, Virginia Carlson, told me near the end of her life (in the late1980's) that their parents had urged my mother to divorce Frank and to make a life for herself sometime in the early 1950's.

That was probably about the time that Dad went into Patton State Hospital in 1948-49. Mama was fully cognizant of the dependent nature of my father, and must have realized that he would have probably gone to pieces on his own and might have joined the homeless somewhere in south San Diego. The ultimate outcome of her decision was that my Dad played the role of *paterfamilias*, always drove us on any car trips, and was to be obeyed for any and all chores around the house. My main jobs consisted of laying the fire in the living room fireplace and changing the pets' water daily. This was good training for later in life.

    My parents were very keen on classical music concerts. I have fond memories of the concerts, which were mostly in the Russ Auditorium, on the edge of the San Diego High School campus. My parents encouraged me to take summer-school courses at San Diego High. I learned to touch-type in the seventh grade, and took some Band and instrumental ensemble courses. These were during the mornings, and then I would go to the YMCA at 7th and C Streets, the old Marston mansion, covering an entire city block. It had a large swimming pool in the basement.

11.
Activities in Junior High school

I worked out with the YMCA swim team in that pool during my Junior High years. I rode the bus to and from the summer school and YMCA and home, and considered myself quite the man about town. Our swim team competed at several YMCAs in southern California. One weekend we drove in an old panel truck to Hollywood for a swim meet. In downtown Hollywood, our coach, Eddie Callahan, leaned out the driver's window and said: "Hi, Freddie!" Fred Astaire was crossing the street in front of us. That was a thrill.

Unlike our neighbors, the Colemans had a variety of English cars, beginning in November 1950 when Dad bought a new Hillman Minx off the lot in downtown Ocean Beach. It was quickly taken over by Mama, as she needed it to commute to and from La Jolla High School, where she taught for four years. Dad made do with the old Chevrolet inherited from the grandparents, and then traded it in for a new Morris Oxford in spring 1952, in time to drive us up to Montana in one of our frequent pilgrimages to the maternal relatives. In hindsight, we were on the tail-end of being out in the wide-open spaces of the West. Going northward, we traveled for hundreds of miles on U.S. highway 395, camping out in fields up along the California and Nevada borders, and then on via Wyoming and Yellowstone National Park. There was very little traffic except for our time in Yellowstone.

Returning home, we took a roundabout route up to Calgary, Alberta and across western Canada on Trans-Canada Route one. In far western Alberta, the road became gravel two-lane, and we followed the Big Bend of the Columbia northward and then southward, seeing perhaps one or two cars in the entire day, and the Northland Greyhound bus roaring by, trailing a large rooster-tail of fine white dust. What a contrast to the thousands of Asian tourists we saw in busy traffic during a trip on wide superhighways in Alberta and B.C. nearly sixty years later! We arrived in Vancouver, took the ferry to Victoria, and visited many Coleman relatives (the great aunts Bertha and Louie and Uncle Leslie at Deep Cove, north of Victoria). The remainder of our trip was down the west coast. We stopped off at Reed College, and dropped by the president's office. A young Duncan Ballantine welcomed us, and showed us around the lovely campus. Little did I realize that I would attend Reed beginning just four years later.

In 1956, I went off to Reed College, and a few weeks later, Dad proudly announced that he had traded in the Hillman on a MG Magnette, a sporty four-door sedan. Mama loved driving it around town for nearly 14 years. The Oxford was not thriving in the heavy traffic of San Diego, so in 1957 Dad traded it in on a used Rover 75, an impressive six-cylinder car. That worked out well except for the frequent failures of its electric fuel pump, nearly annually. Dad and I felt that we had given the English cars a good try, and checked out US made cars in early 1960. After test-driving several economy cars, including the Plymouth Valiant, the Studebaker Lark was our top choice, and Dad got a lovely light green one with a V-8 engine and overdrive. It handled well on several cross-country trips, including the trip in mid-summer 1965, when the family drove east to Philadelphia to my wedding to Frances Evoy, my bride of over 57 years now.

During most of my years in Junior High School, I had a paper route. I began in June of 1950. I vividly recall the day we declared war in Korea, with the screaming banner headline: "Nation at War." I started out with thirty-four subscribers across a more than two-mile route, from Alhambra Street on the north, to Hill Street on the south, Sunset Cliffs Boulevard to the west, and Catalina Street on the east. I folded the papers beforehand, wrapping them with rubber bands when they were thick, and in a neat tucked- in folding technique, when they were thin, that enabled me to sail the papers onto the front porches of my customers. At its inception, I earned between $35 and later up to $60 monthly on the route, as the number of customers nearly doubled over the ca. four years' time. There were many vacant lots that far south on the ocean side of Point Loma at the beginning of the 1950's. They filled in rapidly during the late Truman and early Eisenhower years of what seemed like prosperous times. The totality of the times was not always rosy: In seventh grade band, the instructor, Mr. Ortiz, told us to drop to the floor if we saw a blinding flash outside the windows, indicating the detonation of an atomic bomb had occurred. I naively raised my hand, and asked, "What do we do with our instruments?" He shook his head, grimaced, and didn't respond. Being a Navy town and full of defense industries, San Diego took its role seriously in possibly being a prime target of a surprise Russian attack.

12.
My solo trip to Montana

One month after beginning my paper route, I set out on a trip to Box Elder, Montana, to my uncle and cousin Bill Cowan to work for a month. I boarded a Greyhound bus at San Diego, transferring at Los Angeles, and again at Salt Lake City. The next leg of the trip was at Butte, Montana, where I transferred to an Intermountain bus to Helena, where I spent a week with my Uncle John and Aunt Virginia Carlson, and their ten-year old daughter, Carol Jane. I was nearly twelve years old at the time, and feeling quite independent. I had made friends with Jerry from Hayward on the bus from Los Angeles across the vast western landscape. We listened to radio and jukebox tunes at the many rest stops en route. Foremost among these were *Does the Spearmint Lose Its Flavor on the Bedpost Overnight?*, *Salty Dog Rag*, and *How Much Is That Hound Dog in the Window?* by Homer and Jethro.

  Time with the Carlsons passed rapidly, playing croquet in their shady back yard, and enjoying outdoor picnics. Aunt Virginia and Carol Jane put me aboard another Intermountain bus from Helena in mid-morning, which got me into Box Elder in mid-afternoon. I stayed with my grandmother Cowan. The night freights on the Great Northern line lulled me to sleep on the warm July nights. Uncle Bill had many jobs in mind, and set me up with paint and brushes to completely repaint a long white fence that ran between Uncle Bill's and Grandma's place. There seemed to be an endless number of palings to sand and paint, and I watched the passing parade of cars and trucks and trains over the next three plus weeks. I wore a wide-brimmed straw hat and long-sleeved shirt to avoid sunburn. Cousin Bill and I took occasional afternoon breaks, driven by Aunt Annabel to swim in nearby farm ponds.

  Uncle Bill Cowan expected the best of us, his son Bill and

me, and that was a life-long trait of his. During World War Two he had enlisted in the Army, and wanted to serve either in Europe or in the Pacific. He made the rank of Captain and was put in charge of basic training at Camp Roberts, north of San Luis Obispo on the central California coast. He worked his recruits hard, and was commended for his service in that capacity. When he applied to serve on the front lines, the Army hierarchy said he was too important to the war effort, so kept him "in harness," so to speak, for the duration of the war, much to his dismay.

By early August, my parents and Margery drove up, visited a few days, and took me back home. Thinking back on the timing, what was I thinking, turning over my paper route to a local friend, Bobby Pitha, only one month after beginning that job? The summer break was very welcome, however, and a total contrast to the suburban Southern California life.

13.
Domestic help for our household

During my years in Junior High School, mom and dad hired a series of live-in maids who lived with us during the week and fixed food for us and kept the house clean. They were all Mexican, and obtained through a woman in our neighborhood who arranged for domestics to come to many homes in the San Diego area. They ranged from Rejina, a grandmotherly lady from Tijuana; to Genevieve, a young girl in her early twenties from Ensenada; to Juana, from Jalisco in the main part of Mexico; to Minerva, a fortyish woman from Mexicali. They all lasted from several months to one year, and were very interesting and made helpful comments as my mother and Margery and I tried out our very limited en Espanol on them. Dad already knew a lot of Spanish. The worker program gradually died out, and we changed to having several local ladies, none of whom were nearly as interesting or, it seemed to us, interested in doing a wide-ranging job of cleaning as well as cooking. The main benefit from the hired help was that Mama managed to finish up her Master's degree at San Diego State College. From then on, Margery and I learned to be more helpful where we could in helping to fix the evening meals, and wash dishes, etc. Thus ended a rather bizarre but fascinating look into the world of dealing with hired help. All this was achieved at some cost to the family budget, as mom and dad were just getting their heads above water, financially.

In my sophomore year at Point Loma High School (1953-54), I took a year-long course in Biology. This was something I really wanted to excel in, and I avidly read the textbook and also books on forestry and wildlife. I aspired to be a forester, and figured that would enable me to be outdoors a lot and also to observe the flora and fauna in whatever sites I worked. I wrote up a term paper, called *"Forestry: Vital to the Nation."* I cited Donald Culross Peattie, John Muir and other environmental stalwarts. Upon taking it in to Mr. Farrar, my Biology teacher, he asked me: "What do you want me to do with it?" He had not assigned it, and really preferred not to read it. That was quite a disappointment to me, providing a life lesson that it is best to enjoy what you are doing and not necessarily seek the approval of those in authority.

14.
Doing the Impossible: Travelling from North to South by rail in the western United States

*I*n the summer of 1954, I worked in Montana in my Uncle Bill's general store, grain elevator, and on ranches owned by friends of his. Uncle Bill was eager to have his friends share the burden of mentoring me. One of these men, Cecil Mack, took me on for a week or so to assist in stacking hay bales on his 1000-acre ranch. He was rather easy-going, and happy to offer some advice to a young "city-slicker."

By early August, it was time to plan an itinerary to travel from Box Elder, in north central Montana, to my home in San Diego. I had no travel agent, but relied on the train timetables in the local Great Northern depot, operated by Mr. Harry Widowfield. Harry was in charge of ordering boxcars to ship wheat out of town, and also handled the waybills (routing slips, and freight charges) to be paid to the Great Northern Railway.

It was a real step back into the late nineteenth century to visit the old depot along the tracks on the east end of town. It was always quiet and cool in the two-story wooden depot, with the only sounds being the tick-tick-tick of the telegraph and the tick-tock of the old wall-mounted Seth Thomas clock. There were two freight trains a day, the morning one to Great Falls, and the late evening one to Havre, and the Western Star, a passenger train, passing through in the early 50's at various times, twice a day, to and from Seattle.

On occasional breaks from work or in the early evening, I visited the depot and picked up the many vividly-colored timetables for all of the western railroads, including Santa Fe, Southern Pacific, Union Pacific, Rio Grande, Western Pacific, Great Northern, Northern Pacific, and the third northern transcontinental, the Milwaukee Road. I especially wanted to ride it part way on my trip to the west coast, as it used mostly electric locomotives across central and western Montana to Avery, Idaho, and then across central and western Washington as well. Being an avid railfan, it was almost as much fun planning possible itineraries as it was to make the trip.

The objective was to visit relatives en route, saving on costs of room and board. Uncle Bill teased me about "riding the grub line." Leaving Box Elder, I went with my aunt and uncle Virginia and John Carlson to Helena, spent a day or so there, then rode the Northern Pacific "*Mainstreeter*" to Missoula, where I visited two days with my great-uncle Rufus Coleman and his wife Rhilda. Rufus, a professor of English at the University of Montana, gave me a book on the history of Chief Joseph and the Nez Perce Indians of eastern Washington. It was fascinating. I then rode the Milwaukee's "*Columbian*," with its box-cab electric "motors" to the west. The electric locomotives were very powerful, accelerating the train rapidly, but maintained about 40 miles-per-hour for the duration of the trip. We changed locomotives at Avery, Idaho, with a Diesel GP-9 road switcher taking us the ca. 120 miles to Spokane. This train arrived at midnight. I walked two long blocks to the Great Northern station, and boarded the overnight southern section of the Empire Builder to Portland, Oregon, on the Spokane, Portland & Seattle line. The contrast in quality of the roadbed of the NP and SP&S lines vs. the undulating nature of the Milwaukee Road main was obvious, and a sign of a troubled future for that fascinating railroad

Upon arrival in Portland, Aunt Barbara picked me up and I stayed with my Coleman grandparents at the Audubon Bird Sanctuary on Northwest Cornell Road for a couple of days. Grandfather took people around the sanctuary, giving nature talks. The final southbound leg of my trip was on the famous "Shasta Daylight" of the Southern Pacific Railroad. The Shasta had modern Alco diesel passenger locomotives, painted in the vivid Red and Yellow Daylight colors. The train, 16 cars long and every seat occupied, traveled all day on the route through western Oregon, then around the north, west and southern slopes of beautiful Mount Shasta, and on to the San Francisco Bay area. Because my reserved seat was occupied, I walked through several cars to find an empty seat. My seatmate was Larry Ford, of Ford and Harris, a vaudeville act. He was a pal of Rochester, a character on "The Jack Benny Show," and he regaled me with tales of his vaudeville life.

Through passengers like me transferred at Martinez, CA, to the "Owl," powered by a steam locomotive from Oakland to Bakersfield, then a diesel for the trip into LA. The Owl was an overnight mostly-mail train (it had 22 cars!) that traveled leisurely down the Central Valley and over the Tehachapi mountains all the way to Los Angeles, arriving one hour late, at 11 a.m. The contrast could not have been greater between the Daylight and the Owl, as the cars were old Pullman green "heavyweight" cars, clean but very dowdy in condition. I transferred to the "San Diegan" of the Santa Fe, for the final 120 miles farther south.

Viewed overall, it was a wonderful trip for a railfan, but not a rapid way to travel, even in the mid-twentieth century. The elapsed time from Portland to San Diego was about 33 hours, six hours longer than riding a Greyhound bus, which was more cramped and stuffy. Amtrak provides same train service on the *"Coast Starlight"* from Los Angeles to Seattle, a fine experience.

My travel route was soon to be rendered obsolete, as many long-distance trains were dropped nationwide. This began early on the Milwaukee Road, with the *"Columbian"* terminating in eastern Montana in 1955, and the *"Olympian Hiawatha,"* their premier passenger train, cancelled in 1961. Many other trains were cancelled throughout the 1960's, leaving Amtrak to run a skeletal operation across country from 1971 onward. As of 2022, the Republicans in Congress are trying to abolish all long-distance trains, despite their earning over 95% of their operating costs. The Biden administration is fighting back, planning to renovate the fleet of passenger cars and locomotives as part of the upgrades in infrastructure.

15.
Some international influences at the end of High School

In June 1956, I resigned my weekend position as a carryout bag boy for the Ocean Beach Safeway and signed on with several dozen other young students at Convair Aeronautical, the company my dad worked for during the mid-1950's. Our group was called the *"Foreign Legion,"* as it had many employees hired as draftsmen from all over the world. The dominant foreigners were Northern Irishmen, who came to the USA in the aftermath of the failure of a major aircraft company, Shorts, Ltd., located in Belfast, Northern Ireland. The young men of the Foreign Legion, many from South America as well, were eager to play their national sport, soccer.

I went to a few of their evening games at fields on the bay side of Point Loma. The two teams had intriguing names: the Irishmen were *"The Ulster United,"* and the others from all over became *"The Alien Athletics."* We high school graduates were put to work updating thousands of drawings of former planes produced by Convair. The principal ones were the Convair 440s, similar to the DC-6 airplane. Some of the adults were involved in drawing parts of the fuselage of the aborning Atlas rocket, that was to be produced out at a new assembly plant on Kearny Mesa, several miles northeast of town. Those were boom-times, and I turned my back on them to pursue a lifelong goal of being a biologist in undergraduate and graduate schools.

16.
Origins of my life-long interests in Biology

*M*y interests in Biology were stimulated at an early age on walks with my parents in the Eucalyptus woods two miles south of us and in visits to the intertidal zones on Point Loma. A chance to study animal life was always encouraged by my father, Francis Coleman, who almost majored in biology at Reed. I enjoyed digging out trap-door spiders in road embankments in the woods. I tried keeping them as pets in coffee cans in my bedroom.

Surprisingly, my grandfather, Norman Coleman, who was an English professor at Reed (and its president from 1925-34) and later at Lewis and Clark College in Portland, had been a Biology major at the University of Toronto, graduating in 1899, complete with a Governor's Gold Medal. He found that he could get a Master's in English (at Harvard) more quickly than a Biology Master's, thus enabling him to marry his lifelong sweetheart, Ethel Fleming, so he went into English literature as a life path. I became entranced by this "third time is a charm" motif, and chose Biology as a major and never strayed from that as an objective

Biology was taught very much as a physiological and biochemical science at Reed. I enjoyed the Humanities and Science courses at Reed, but did not swat at the books hard enough to earn above-average grades. There was so much to read, and so many term papers to write that I, at times, took refuge in my extracurricular musical activities. During my four years there, 1956-60, the years of President Richard Sullivan's early tenure, Reed was barely surviving financially, as was usual. Dorothy Babson McCall, sister of Governor Tom McCall, had donated funds to be used for subsidizing instrumental music lessons for Reed students. I took advantage of that and had weekly lessons with Marge Smith, principal bassoonist in the Portland Symphony, in rooms in the old President's House on the edge of the Reed campus.

Frederic Rothchild, a very gifted pianist, organized many chamber music performances on campus, and by his example, encouraged us to offer concerts, that we termed "Sound Experiments." In my freshman year, we (Herschel Snodgrass, clarinet; Loline Hathaway, Oboe; a fellow from Portland symphony on French Horn; myself on bassoon and Mr. Rothchild) practiced and performed a repertoire of Piano Quintets by Mozart and Beethoven, which we played first for the Eastmoreland Women's Club, and then in the Reed Chapel. Reed had been the "beneficiary" of an old high-mileage 1949 Buick Roadmaster sedan. Mrs. Sullivan chauffeured us over and back from the Women's Club concert, and our entire group, instruments and all (minus Fred and his piano), fit into its capacious seats. The car had only ¼ tank of gasoline, and it was on "drop dead" empty by the end of the ca. 8 mile round trip, causing Mrs. Sullivan (and us) considerable anxiety. A week or so later, we performed the Quintets in the Reed Chapel, as one of the series of "Sound Experiments." Needless to say, the concert in the Chapel was much more enthusiastically received!

Other musical events that were high points in my Reed years included the wonderful Christmas concerts (usually Handel Oratorios) and the Gilbert and Sullivan Operettas in spring, led by Herbert Gladstone. It was truly an opera and concert orchestra of amateurs, playing "for the love of it." One really humorous moment stands out from the others. It was during a performance of "Patience" in the old Botsford Auditorium. Fred White, the lead actor playing Bunthorne, had a line: "What was that?" in response to some foolish riposte on the stage. Just before he uttered the line, one of the woodwinds played several wrong notes, and Fred, without missing a beat, looked out into the orchestra, and said, "What was THAT?" It brought down the house.

In my college years, a few faculty members were standouts, but only a smattering. At Reed College, our Freshman Humanities course had seven credit hours, or about half of an average course load. Four lectures and three conferences, or discussion sessions were held per week. In essence, it was an extended "Great Books" course, from Homer's *The Odyssey* up to the reign of Louis XIV, the Sun King. Warren Susman, an historian, gave brilliant lectures on the Persian Empire. Susman made the emphatic declaration that "history is selective," and how much that affects all historians, and their readers. This sentiment was not original with Dr. Susman, of course, but it made a big impression on my naïve outlook. Susman also presented a most thought-provoking exposition on Cervantes' *Don Quixote*.

Another influential Reed professor, again in the liberal arts field, was the charismatic Lloyd Reynolds, our resident Calligrapher and teacher of a large survey course called "History of Art." He made use of ancient audio-Visual equipment, including an epidiascope, a device for showing pictures in numerous books. He famously interpreted art and historical phenomena in general viewed through the lens of Zen Buddhism. This is not easy to do with western art, but he carried it off with both aplomb and panache. He loved to introduce feeling via such comments about art subjects "making themselves heavy," and other bon mots. One of his favorite devices was to say a Zen koan, or saying. One of my favorites is: "Why is a mouse when it spins?" That was quite a contrast to the Socratic dialogues in the introductory Humanities course.

Often on Sunday mornings, I met my grandfather, Norman Coleman, at the First Congregational Church in downtown Portland. After attending the service, we visited with people in the coffee hour, and on every other Sunday, Grandfather and I hitched a ride to their home in the hills above northwest Portland and had Sunday dinner with them. Because I was fond of both grandparents, it was a welcome break from the incessant need to write reports and term papers at Reed and have a family visit. During the long Thanksgiving weekend, Aunt Barbara invited me over to have dinner with her two children, Carolyn and Debbie, and the grandparents as well.

As a Biology major, taking many courses in science and math, I hoped to encounter notable professors in those courses. I was impressed by the superb drawings of developing embryos made by Dr. Larry Ruben, but little philosophical insights accompanied them. I was not a stellar student, earning a C average, writing seemingly innumerable term papers, and not impressing the biology or chemistry profs.

In occasional chats with Dr. Les Squier, Dean of Men at Reed, I gained some insights into things to strive for in life. He commented on a few students at the college who had "original ideas," and valued their insights. This aspect of my life seemed greatly lacking, and I valued his perspective. I gained valuable insights from Robert Ornduff, the visiting botany prof. in my junior year, who taught the Plant Evolution course and Frank Gwilliam, my adviser and teacher of the Invertebrate Zoology course in my senior year. Ornduff discussed at great length the controversial ideas of Alfred Wegener, who espoused the idea that continents drifted over time. Many geographers vilified him for his unorthodox ideas, and at the time I took the Plant Evolution course (spring of 1959), there was no known mechanism for what became known as "Continental Drift" just a few years later. Ornduff also emphatically proclaimed: "species is a concept of man," and not a stand-alone entity. Dr. Tahir Rizki, our genetics professor, strongly disagreed with Ornduff, but current views on species show the validity of Ornduff's comments.

In his Invertebrate Zoology course in my senior year, Gwilliam commented on Pogonophora, strange worms without complete digestive tracts (which Gwilliam quaintly phrased as "they had no ass") retrieved from deep-sea trawl samples in mid-ocean by Russian scientists in the early 20th century. Decades later these creatures were studied via deep-sea submersibles and identified as Vestimentiferan worms, part of chemosynthetic foodwebs at deep-sea vents in mid-ocean. This was an interesting early insight for us nascent biologists.

At a similar time, I also recall my father and Ed Dacus, two physicists with Ph.D. degrees, and their predictions about the future success of fusion technology for use in generation of electric power. Ed worked at General Atomic, a fusion research facility in suburban San Diego in the 1960's. In 1960, they confidently predicted that all of the technical problems with fusion would be worked out in the next thirty years. A full 62 years after their prediction, the field seems as far away from a solution as ever.

With a lifelong interest in the outdoors and forestry, I applied to Larry Thorpe, District Ranger of the Canyon Creek District in the Gifford Pinchot National Forest in southwest Washington State for summer work at the end of my freshman year at Reed. Mr. Thorpe was a friend of Carleton Whitehead, the director of Alumni Relations at Reed, who had suggested that I apply for a student trainee position for the summer.

I did mostly grounds maintenance and general handy-man work the first summer (in 1957) and then in the summer of 1958 I worked on an "engineer" survey gang, who surveyed access roads into timberland to the north and east of our ranger station. Our work entailed getting compass bearings and measuring distances to the nearest tenth of a foot. This work was more interesting, as we certainly hiked enough miles (occasionally up to six miles round trip) into and out of our survey sites that kept me in the best physical shape I have ever been in. There was virtually no science involved in the work, so during my junior year at Reed (1958-59) I decided to pursue academic research. I applied to be an NSF-funded summer undergraduate research participant (URP) at the Scripps Institute of Oceanography at La Jolla, ten miles up the coast from where my parents lived in Ocean Beach.

This work involved maintaining cultures of marine bacteria and using ultraviolet light to make auxotrophic mutants (strains that were unable to synthesize one or more amino acids, in this case) of *Serratia marinorubra* to bioassay seawater for the presence of amino acids. The work was really interesting; I made time to visit the Scripps library and read up on aspects of marine biology, about which I knew very little. I talked with Drs. Bill Belser and Phyllis Bear, the senior scientists in the group, and learned a lot from them in the process. All the Petri dishes and pipettes that we used were Pyrex glass, and the amount of autoclaving and washing that we did was very great indeed, in that age decades before plastic disposable labware came into vogue. Armed with my knowledge of the bioassay procedure, I took some cultures of *Serratia* with me to Reed College to measure the amino acids in the stream that ran through campus on my research thesis for the Bachelor of Arts degree. This was a life-changing moment for me, as it moved me into a life of academic research, and away from the applied ecology life in Forestry, although I returned to ecosystem research in forests later on in my career at the University of Georgia.

The administrators at Scripps Institution of Oceanography had so many applicants for the thirty or so undergraduate research positions available (at $300 for three summer months) they split the stipends into thirds, and hired more than one hundred students, paying us $100 each for the summer experience. We were delighted to have the chance to work in such a famous place, and to see, if not talk with, people like the Nobel laureate Harold Urey (the discoverer of the heavy hydrogen isotope Deuterium) and the world-famous ichthyologist, Carl Hubbs. Stanley Miller, a recent student of Urey's, a few years earlier had carried out the Miller and Urey experiment on the origin of life from the interaction of simulated lightning and inorganic salts in seawater, proving that organic compounds could arise thereby, and adding to the excitement about the biochemical basis of the origins of life.

Dr. Roger Revelle, director of Scripps, a bronzed six feet five inches tall extrovert, gave the URPs an unforgettable introduction to Scripps covering studies in biology, chemistry and of course oceanography. He had a pithy phrase that he emphasized repeatedly:" a successful scientist has a seeing eye and an inquiring mind." This was heady stuff to us, and we were hooked on doing science.

My biology adviser, Dr. Frank Gwilliam, an invertebrate zoologist who came to Reed in the fall of 1957, spent his summer doing research at Scripps simultaneously with my sojourn there in 1959. Upon being informed by Dr. Belser that I was doing a first-rate job of research in Dr. Belser's laboratory, Gwilliam urged him to write a letter for my file in the Biology Department at Reed, as my academic progress there had been less than stellar. My overall grade record, which I found out only after graduation (Reed wanted no competition for grades among its students) was only a C- average, reflected the difficulty of the courses. For example, the first midterm in Freshman Physics, which I took in my senior year, had an average of only 30 points out of 100. Expectations were high at Reed in the 1950's.

Moving ahead into my senior year at Reed, armed with the techniques I learned at Scripps, I got some useful data on amino acids in the Reed Canyon Stream from water samples that I took from the stream, using the auxotrophic mutants from Dr. Belser. I worked up many tables and figures with advice from Dr. Tahir Rizki, my thesis adviser, who was a geneticist with extensive knowledge of the use of bacterial cultures in biological research. In February of 1960, Rizki phoned Dr. Peter W. Frank, an aquatic ecologist at the University of Oregon in Eugene, urging him to hire me as a pre-doctoral research student. Rizki had met the famous G. Evelyn Hutchinson at Yale University a few weeks before coming out to his teaching job at Reed in 1956, and was urged to keep his eyes on a bright young former postdoctoral of Hutchinson's, Dr. Peter Frank, who might be a good person to place a Reed Biology graduate with.

17.
My grad school years at the U. of Oregon

I drove down to meet Dr. Frank with my friend Ray Swanson, talked matters over, and was surprised and pleased to be offered a research assistantship on the spot. I had been accepted at only one graduate school, the University of Chicago, which welcomed many Reed graduates. I opted for the University of Oregon, accepted on probation due to my low grades, as the milieu sounded better for me, with the chance to work on a research grant as a research assistant.

In the years from 1960-64, I obtained my M.A. and Ph.D. in Biology with Peter Frank as major professor. Dr. Frank had obtained his Ph.D. in 1951 with Thomas Park at the University of Chicago. After he spent a postdoctoral year with G. Evelyn Hutchinson at Yale, and a five-year stint at the University of Missouri, he moved to the U. of O. in 1957. Frank emphasized the demographic aspects of field biology, building life tables of various invertebrates—Daphnia in the laboratory, and limpets (*Acmaea* spp.) and sea urchins (*Strongylocentrotus purpuratus*, with his graduate student Tom Ebert) in the field. In his introductory ecology course, Frank used the second edition of E.P. Odum, Fundamentals of Ecology (1959) as a text, and encouraged us to view the natural world from an ecosystem perspective, as well as the more mathematical, demographic worldview that pervaded his thinking.

The postdoctoral experience Dr. Frank had with Hutchinson influenced how Frank addressed ecological problems, and his lectures included many examples that Hutchinson had used in his lectures on ecology and limnology (the study of lakes) at Yale University. Hutchinson was a true polymath. He wrote many ground-breaking papers, including the famous "*Homage to Santa Rosalia*, or Why are there so many species of Animals?" He wrote articles that appeared regularly in the Marginalia section of The American Scientist, which I found captivating. My father subscribed to that journal at home during my high school and college years. Such articles as *"Copepodology for the Ornithologist"* and *"The Cream in the Gooseberry Fool"* were gems of biological and ecological insight. A few years later, Hutchinson wrote a seminal paper in a 1970 issue of Scientific American, entitled: "The World as a Heat Engine." Such an intellect, and such a wide scope of topics! Hutchinson, along with Amyan Macfadyen, was one of the true giants to appear in the field of Ecology.

For my Ph.D., I studied the laboratory population ecology of ciliates (large protozoa) of the genus *Kerona*, which were epizoic on the surface of freshwater *Hydra* species in Triangle Lake, ca. 30 miles west of Eugene. I obtained cultures of several species of Hydra from Dr. Helen Park in the National Institutes of Health, and measured how rapidly the ciliates increased over time on the various hydras. It was all measured carefully in replicated experiments, with little thought given to what was going on in the "real world" outside of the laboratory. I never did elucidate what factor(s) made the ciliates with their hypotrichous cirri (large modified cilia on the bottom surface of the ciliates) so successful on the hydras. I got my Master's and Ph.D. degrees within four years, graduating in August 1964. I published one major paper from my dissertation (Coleman, 1966).

My first two years at the U. of Oregon were supported by a Research Assistantship, starting out at $1700 per year. I worked with Dr. Frank on his NSF-funded project on the population dynamics of limpets (primitive snail relatives) at the Oregon coast, at the Oregon Institute of Marine Biology. Frank had a group of summer undergraduate research students, working on funding from the NSF similar to the stipend I had received in my summers at Scripps, whom I supervised on the project. We tagged over four thousand limpets individually with numerical labels. There was a steady loss of limpets over time. We retrieved the shells and noted the mortality, but never made the connection to the fact that oystercatchers (birds) had pried them off and eaten their succulent insides.

The Institute was a group of old wooden buildings built by the Civilian Conservation Corps (CCC) in the mid-1930's. The buildings served their purpose after a fashion, but required continual maintenance. Our resident handy-man was Harry C. ("Doc") Luce. He had an amazingly varied assortment of plumbing parts, miscellaneous lumber and odds-and-ends. Peter Frank, who was the Institute Director, was consulting with him almost daily about needed repairs. Doc was very patient, but had to be coaxed along at his own pace. He would climb in his old 1941 Dodge flatbed truck and roar off to fix whatever needed attention. Our station had salt-water aquaria; Doc had to keep a pumping and filtration system in operation, which was a major concern for him. As time went by, Doc felt lonely, and prevailed on us grad. students to write to his recently-widowed sister-in-law, Mary, back in upstate New York, urging her to come out to Oregon. Mary came out in my second year, 1962, and kept him company. Doc passed away two years later. We lost a colorful and devoted character, who couldn't be replaced.

A major life-path altering decision was one I made early on in my graduate school career. I had never had a course in Calculus during my years at Reed College. I duly enrolled in an introductory Calculus class at the University of Oregon during the fall term in 1960. I ran into difficulties early in the course, and didn't get any tutoring assistance, perhaps because I was too proud to do so. As a result, I failed the final exam and that record remained on my graduate school transcript from then onward. Not surprisingly, when I finished my Ph.D. dissertation and graduated from the U. of Oregon, I got no responses to applications for postdoctoral stipends or openings. This development was one that had some long-range impacts, and helped give me the impetus to change fields to a very different discipline within ecology. This career move was one that ended up paying major dividends for me over the ensuing decades.

It was apparent that my primary research interests were somewhere else than in freshwater, and I decided to branch out into soil ecology. The ecology of decomposition processes, a topic on which I had written a term paper for Dr. Frank, and gotten an "A" grade from him in my first year at the U. of O., held great fascination for me. When I received my Ph.D. degree in 1964, Peter Frank asked me what I planned to do next. I replied that soils and their ecologies were a great unknown. He said there was a reason for that, as soils are inherently "messy." He meant that one couldn't see into them, and hence were not amenable to definitive studies. That was a real challenge to me, and in my subsequent career, I was determined to prove him wrong. Soils may be opaque, but they are structured and not messy. I discuss below my life-altering interactions with Prof. Amyan Macfadyen, with whom I conducted my post-doctoral research in Great Britain.

18.
Comments on the supportive nature of my parents

By the late 1950's, my father had a few more episodes in his life history of bipolar problems. After an extensive recuperation in the fall of 1958 in a sanitarium in Chula Vista, south of San Diego, the attending physician prescribed a mood-stabilizing drug, called Melloril, and Francis was able to more or less maintain his life on an even keel. The events leading up to this episode were traumatic for my mother and sister Margery and me. Alice's comment was: "If we can't get your dad straightened around, you may have to forget that you have a father." With these uncertain words echoing in my conscience, I returned to Reed for my last two years of class work.

A few years later, at the time of my travel to Europe to do my postdoctoral work with Prof. Macfadyen, my mother balked at providing me with travel money and expenses for yet another year of what seemed like a prolonged doctoral study time. Fortunately, Dad stuck up for me and said, "David needs to have the postdoctoral experience. It will make him more attractive as a possible academic hire." He was correct in that assessment. As we will see in the sections below, this was the chance of a lifetime for me to branch out into an area of study that I pursued successfully for more than forty years.

19.
Post-doctoral Experiences in the U.K.

I had read several papers by Amyan Macfadyen, including a review of the ecology of soil microarthropods in volume 1 (1962) of *Advances in Ecological Research*, and his ecology textbook, *Animal Ecology: Aims and Methods* (1962). In the spring of 1964, I naively wrote to see if I could work with him on a postdoctoral fellowship. Macfadyen wrote back with bad news and good news: the bad news was he didn't have any funding, but the good news was he had ample laboratory space. In the end, he found some Teaching Assistantship funding for me to assist in laboratory and field courses that he was teaching at the University College of Swansea, in South Wales. Armed with his letters of encouragement, I packed my bags and set off for Europe in late August 1964.

Amyan Macfadyen (left), David Coleman (center) and colleague in the garden behind the Zoology Department, University College of Swansea, Wales, U.K., June 1965.

This was after being declared 1-A by my draft board. I had more than a few anxious moments meeting with the draft board in downtown San Diego. I had no objections to being drafted, but had this one chance in a lifetime to work with the noted Prof. Macfadyen. They seemed quite informal about it, and one man on the board, sensing my anxiety, said: "Oh, let him go."

Determined to see *Europe on Five Dollars a Day*, the title of a travel book of the early 1960's, I took the Leonardo da Vinci of the Italian Line on the eight day voyage from New York to Naples. Traveling northward by train, I did some sightseeing in Rome and Florence (seeing art treasures in the Pitti Palace), economizing by staying in dormitories and pensiones. I then traveled westward on an overnight train to Paris to the Hotel Raspail, where I met Frances Evoy, who would become my future bride-to-be. We had a memorable 3-4 days of sightseeing in the early September heat. We had met originally during a summer break that Frances spent with her brother Bill in Eugene, Oregon in 1963. I was so impressed by her that I sent, that Fall, an LP recording of Schubert's "Trout" quintet, in hopes that our paths might cross again

High points of our time in Paris included a day trip out to Versailles, eating croque monsieur sandwiches in brasseries, and riding the Metro around town. One of our favorite museums was the Musee Rodin, which had numerous life-sized bronzes of men and women. I was particularly impressed with his marble sculpture of the woman with the cascading hair. It rivaled Michelangelo's beautiful statue of David, which I saw in the Pitti Palace in Florence en route to Paris.

From Paris I went to Calais by train, took the channel boat to Folkestone, then on by train to London. The influence of Macfadyen again came to the rescue on the boat trip across the English Channel. Customs and Immigration officials called all the passengers in, individually, to be interviewed to give their reasons for coming into the U.K. My plan to work on a postdoctoral year with Prof. Macfadyen left them very dubious. They would okay a tourist visit of less than 30 days, but for any longer time period they needed documentary proof. I went back to my luggage, retrieved Macfadyen's letters, and showed them to the officials. After a more than 20- minute wait, they reluctantly allowed me into the country for a longer time period. In London, I took two days to walk around the central city, visited museums, especially liking the Tate Modern, and the Alberto Giacometti sculptures in particular, I also enjoyed the sights along the Mall and the lovely area around Hyde Park. The final leg of my trip went from Paddington Station in west London to Swansea in South Wales.

My choice to work with Macfadyen was truly a life-altering one, because he gave me complete freedom to develop a research topic. He was readily available to discuss ideas. To get to and from the field site, some 15 miles west of the University campus, I purchased an old 1952 Austin A-70 ("Hereford") car for 55 pounds, or about 150 dollars. After some trial forays out on *Cefn Bryn* (Welsh for "ridge hill") above Macfadyen's home in Reynoldston on the Gower Peninsula, I settled on studies of food chains in soils in an indirect fashion, measuring microarthropod recolonization of sterilized soil cores that I inoculated with species of litter and soil fungi that were prevalent on the bracken grazing land.

To better orientate myself, Macfadyen had suggested that I write to Dr. O.W. ("Bill") Heal at the Merlewood Research Station for Terrestrial Ecology at Grange-over-Sands, Cumbria, in northwest England. Heal responded immediately and urged me to come up "for a natter" (chat). With my ailing Austin Hereford car in the shop, I traveled by train to Newport, Wales, then up to Crewe, then Lancaster, and finally out to Grange-over-Sands, with the final leg of the day-long trip being behind a steam locomotive! Such bliss for a railfan. Bill picked me up, set me up in a bed and breakfast, then on the following day, took me to the Institute of Terrestrial Ecology (ITE) lab for a daylong visit. Bill and I had interests in protozoa in common, so we compared notes. I also met a real gem of a mycologist, Dr. Juliet Frankland, who later provided several cultures of litter and soil fungi. The insights I got from Bill and Juliet were very illuminating concerning ways to study soil systems, particularly using fungi in food-chain studies.

I proceeded to design an experiment using soil cores from *Cefn Bryn*, a grazing land site on the Gower Peninsula dominated by ferns (*Pteridium aquilinum*) and a grass (*Nardus stricta*), following the recolonization of the cores by microarthropods. The cores (5 cm. dia. x 10 cm deep) were sterilized with between 2.5 and 5 MegaGrays of gamma radiation from an industrial 60Co source at the Atomic Energy Research Authority facility in Wantage, Berkshire, near Oxford. I then inoculated the cores with single species of litter fungi, and after a few days' incubation at room temperature, replaced them back in their original holes in the field. The radiation treatment had a marked effect on the soil cores: there was a strong ammoniacal odor from the impact of the gamma rays on nitrogenous compounds bound to the clay lattices in the soil. However, the roots and plants seemed to recover readily upon being placed back into the holes they had been removed from.

After a few weeks in the field, the cores were retrieved and extracted on Tullgren (high-efficiency thermal gradient) funnels. We found that certain mites and Collembola, as inferred by their recolonization patterns, exhibited significant feeding preferences for fungi in the field (Coleman and Macfadyen, 1966). That was novel then; many prior preference studies had been conducted only in the laboratory.

Getting housing, or "digs," as they are called in the U.K., was no easy matter in those early post-war years. While I was still corresponding with him from Oregon, Macfadyen had suggested a person from whom I could rent a room but I couldn't be bothered to plan so far ahead. Once I arrived there, I answered numerous ads in the South Wales Evening Post, but found them either very depressing or too costly for my flat wallet. Mrs. Hoyland, Professor Knight-Jones' principal secretary, urged me to place an ad in the local newspaper, noting that I was an American postgraduate student, looking for accommodations.

I got several phone calls, and the one that intrigued me the most was from Mrs. Mary Hurley, living in a home in "the Mumbles," a suburb west of campus, perched on a high hill with hundreds of other multiple-unit homes. She had a basement flat with bedroom, kitchen and bath, which she offered for three pounds and six shillings per week. This translated into about nine dollars per week, which was within my budget. Even more importantly, she and her daughters, Catherine and Mary, took turns inviting me upstairs on weekends to have afternoon "tea." This consisted of large sandwiches, cakes or pie of some sort, all washed down with copious amounts of good hot tea. This also made up for the generally cold and humid climate in my little apartment. I often stayed on campus for dinner in the Junior Commons Room, where I could get a decent meal for about one shilling and sixpence, roughly equivalent to twenty-six cents in USA currency. The pound, at that point, was the equivalent of $2.70. I worked often on into the evenings in Macfadyen's large, well-lit and warm laboratory on either field notes or writing up my dissertation for publication.

The social milieu on campus in the Department of Zoology was quite genteel. The professor, Ellis Wynn Knight-Jones, was a noted marine biologist and raconteur. He insisted that staff and graduate students and postdoctorals plan to have morning coffee and afternoon tea in the social room just opposite his office. He liked to tell anecdotes and would occasionally ask others to volunteer topics of interest to the group. We usually numbered 30-40 people at a peak of participants, and it was a great way for me to get to know people. There were times when I wondered if I had landed at a college at Oxford or Cambridge. Professor Knight-Jones, who was proudly Welsh, had a very mellifluous voice and very British accent. If he liked someone's joke or comment, he would chime in with: "Oh, bally good!" This was a far cry from the west coast of the USA, indeed.

The span from September through early December was very

eventful for me personally, as I had really fallen in love with Frances Evoy. We met two times in London when she flew over to visit with me, as we went sightseeing around the area. She also flew over once to Cardiff and we were able to see some of the countryside in and near Wales. We drove on a misty afternoon to Tintern, where the famous Tintern Abbey is located (insert photo).

On Thanksgiving weekend, not observed in Great Britain, Frank and Marge Gwilliam tracked me down via the Reed alumni office and invited me to their home in Bristol, where Gwilliam was on a sabbatical year at the University of Bristol. I rode British Rail over from Swansea; Frank met me at the large Temple Meads station. They had a big turkey dinner, with several other recent Reed grads included, and it was a very welcome touch of home and friends.

By early December, Fran's father's health was failing health due to cancer, and she needed to cut short her year abroad at the Sorbonne. She booked a flight home in mid-December. The weekend before her departure, I flew from Cardiff to Le Bourget airport in Paris. We spent one evening and early morning saying our farewells. At the end of that evening, I proposed marriage to her, and was delighted, and a bit surprised, that she consented. All of this was so sudden that I had no engagement ring, but ordered one from a jeweler in Swansea and sent it to her in February. Her father passed away in early January. It was an eight-month wait until we were wed in August, but it seemed like a lifetime. We wrote copious letters back and forth; mine were daily. We had a chance for a further visit during one week in June. I discuss that later on. We have been happily married for over 57 years now.

Joining Amyan Macfadyen and his colleagues during my postdoctoral year of 1964-1965 marked a sea change in my professional activities. Amyan (insert photo) was one of the pioneers of Ecosystem Studies in Great Britain and in Europe (Coleman, 2010). He was one of the founders of the International Biological Program (IBP), which had a strong British component to it. Macfadyen was interested in all aspects of production ecology. He encouraged me to pursue investigations of detrital food webs in soils. Amyan made his library of over three thousand reprints available to me, and I read eagerly through many of them. Many of the reprints had been used by him as references for his textbook, entitled: *Animal Ecology: Aims and Methods* (Macfadyen, 1962). This book and Amyan's numerous scientific papers were very influential on ecosystems researchers in the University of Georgia group that I joined in September 1965, at the Savannah River Ecology Laboratory (SREL) in Aiken, South Carolina. My postdoctoral year was advantageous in more ways than one, because Macfadyen was very highly thought of by Gene Odum, Frank Golley and others at the University of Georgia. Upon applying to them for a postdoctoral position, they strongly encouraged me to come to work with their group at SREL.

Eugene P. Odum in his office, ca. 1970, at the University of Georgia, Athens.
(Photo courtesy UGA Odum School of Ecology)

20.
Further courtship during one week in June 1985

*F*ran's mother very thoughtfully advanced her the funds to fly over to have a short holiday with me so we could get to know one another better. We went pony trekking on the Gower Peninsula and then took several days to visit Ireland. We drove to Fishguard, in West Wales and took a boat over to Rosslare Harbor, hoping to take a train to Dublin. The railways were on strike that week, so we hitchhiked our way up, bags and all, and got to meet some colorful Irish characters en route. We found a tourist hotel in Dublin, and rented a car there to travel farther north into County Clare. We saw the awesome 700 foot high Cliffs of Moher (insert picture) and reveled in the sunny Irish weather.

Upon our return to Mumbles, we got a stamp of approval, as it were, from Mrs. Hurley and her daughters, visiting over tea. It was difficult to take Franny to the airport and endure another six weeks before returning home for good in mid-July. We were married one month later at St. Thomas' Church in Whitemarsh, just outside of Germantown, PA.

21.
Early years as an academic researcher

After our marriage in mid-August 1965, I took my lovely young bride Frances to the horsey environment of Aiken (race horses overwintered there), and began a life at the Savannah River Ecology Laboratory (SREL) as an ecosystem ecologist and researcher and teacher in the field. We purchased a small three-bedroom home across from a cotton field. The house cost $14,250, and the mortgage payments, at 4.5% interest, were a bit over $91 per month, including taxes and insurance. We could even afford a new two-door Ford Falcon early in 1966, which cost all of $1980, including taxes. My annual salary on the postdoctoral was $9,000, before taxes. Life was good indeed.

I pursued several lines of research at SREL, beginning by using radionuclides to tag food items in the field, to further study microbial and faunal food webs. To gain the information and expertise necessary to carry out this work, Dick Wiegert and Frank Golley at the SREL encouraged me to go to "the source of knowledge," D.A. ("Dac") Crossley, Jr., in the Environmental Sciences Division at Oak Ridge National Laboratory, Oak Ridge, Tennessee.

Dac invited me to visit him in October 1965 and spend 3-4 days on a short tutorial concerning the mathematical calculations of biological half-lives, culture methods for microarthropods, etc. He gave me an overview of what they covered in a two-week short course in four days. It was an eye-opener, how much could be done by such a talented and focused person, all of this presented with a wonderfully dry sense of humor. I also met Jerry Olson, Dave Reichle, Stan Auerbach (head of the Division), and several other colleagues. George Van Dyne was also there, but unavailable that week. I had numerous interactions with him later on in my career.

Returning to SREL, I consulted further with Gene Odum, who

persuaded me that using the gamma-emitting radioisotope of Zinc, Zinc-65, would be the most efficacious way to trace carbon through food webs in the soil, as it was strongly bound to carbonaceous compounds. I performed some trial experiments using an old gamma detector and then, convinced of the need to use an automated counting system to count the radioactivity of hundreds of samples of soil microarthropods over long time periods, I discussed the situation with Frank Golley. As head of SREL, he had access to some Atomic Energy Commission year-end funding, and proceeded to authorize purchase of a 256-channel pulse-height analyzer, complete with automatic sample changer (total cost: $15,000), for the samples of microarthropods I would be counting.

All this was quite a heady experience for someone who had been used to minimal equipment, apart from microscopes. I was allowed to set aside a separate room in a laboratory at SREL, to safely handle the radioactive materials. I erected a barrier of lead bricks on a lab bench, and inoculated flasks with liquid nutrient medium and species of fungal mycelia to which I added large quantities (milliCuries) of the Zinc-65. I proceeded to set up field plots and followed the movement of the isotope from the fungal hyphae into fungal-feeding mites and then into acarine predators that fed upon them. Because of the small biomass of individual species of microarthropod (a few micrograms apiece) we used 99-minute counts per sample. This led to a methods paper (Coleman, 1968) and one in Oikos (Coleman and McGinnis, 1970), as well as a chance to attend an IBP symposium on soil processes held in Paris, at the UNESCO headquarters in November 1967.

At the meeting I met for the first time such famous soil ecologists as Dennis Parkinson and Eldor Paul, and of course my old friends from Great Britain, Clive Edwards, Amyan Macfadyen, and my compatriot from his visit to Swansea in the winter of 1965, the Belgian acarologist and statistician, Paul Berthet.

I digress from this chronological account to discuss the milieu of SREL, and the Institute of Ecology at the University of Georgia. In essence, SREL was viewed by Gene Odum as a sort of "farm team" to send rising young ecologists to the main campus. Those were heady times, with first Dick Wiegert going up to join the Department of Zoology in 1966, followed by Frank Golley in 1967, and Carl Monk to Botany in the same year. Carl was being wooed by Cornell University at that time, and when Carl chose to go to Athens, Cornell hired Robert Whittaker, so we knew Gene Odum (insert photo) and the departments on campus were building a star-studded cast!

In the early years, we traveled from SREL to the campus at Athens at least once a month, driving the 130-mile trip over state roads before Interstate 20 opened up across the Piedmont region of Georgia. We would check out references in the Science library (then in the basement of the Geology and Geography building), and usually have lunch with Gene Odum in the Student Center. Often there would be a late afternoon monthly meeting of the newly-formed Institute of Ecology either on the seventh floor of the Biosciences building or from fall 1966 onward, over in the "Rock House," which was the Lumpkin family's home when UGA was mostly farmland, back in the 1820's.

Gene Odum was the driving force in these meetings, always trying out his ideas and thoughts about moving ecosystem science forward. It was hard to convince him of the logic of one's own ideas, but in future meetings, if he liked what you had said he would bring it up, rephrasing it in his own unique way, and make it much more intriguing and worthy of consideration. Gene was very much in an empire-building mode, but in a novel fashion, constructing what he called his "T model" of collaborations between scientists across departments and colleges to garner interdisciplinary research funding from federal agencies.

Odum was instrumental in bringing Phil Johnson to UGA to help get new research going, including IBP Deciduous Forest Biome funding for work at the U.S. Forest Service Coweeta Hydrologic Laboratory in North Carolina, just across the Georgia border.

Phil left soon afterward to join the National Science Foundation as a program officer and Dac Crossley and Carl Monk took the lead on campus, with Wayne Swank as a co-principal investigator at Coweeta. This began the multi-decade successful hydrology and nutrient cycling studies that carried on into the Long-Term Ecological Research (LTER) era (Coleman, 2010). That funding began in 1980, and went into its sixth renewal proposal funding segment in 2014. However, it is no longer funded.

Shortly before Dick Wiegert moved to Athens in 1966, he urged me to accompany him in April to work at the Atomic Energy Commission-funded Ecosystem study site at El Verde, Puerto Rico. Dick had begun work with Howard T. (H.T.), Gene's younger brother, a few years earlier, on termites (*Nasutitermes costalis*) at the radiation and control sites of the Luquillo rain forest. Dick suggested that I set up a short-term study project with him. Frank Golley readily agreed to provide travel funds. I sampled the biomass of nematodes in the litter and soil of the study sites, which resulted in a two-page paper in the Tropical Rain Forest volume (Coleman, 1971), edited by H.T. Odum. Several other scientists were at the site, including Martin Witkamp from Oak Ridge, whom I had met briefly one year earlier. Martin set up numerous experiments out in the field, and worked long hours. Rather than talking shop in the evenings, Martin preferred to make his lethally strong Mai Tai cocktails, and soak up the tropical atmosphere.

Gene and his brother H.T. Odum (right) discuss concepts in the University of Georgia Institute of Ecology, ca. 1980.
(Photo courtesy UGA Odum School of Ecology)

Martin's papers on decomposition and ways to quantify microbial activity in soils (Witkamp and van der Drift, 1961; Witkamp, 1971) are some of the most insightful in the field of decomposition ecology.

H.T. Odum (insert EP and HT Odum picture) was quite an experience to encounter. He made his brother Gene seem shy and retiring in comparison. H.T. was always on the move, trying out ideas as fast as he could talk, and generally leaving us breathless trying to keep up with the flow of his ideas. One evening on our way back from dinner down on the beach at Luquillo, we stopped at a roadside market for cool beers. Dick Wiegert and I, no slouches in beer drinking, were impressed at the way in which H.T. opened a 10 oz. can of local beer, draining it dry in one gulp. Years later in 1981, the Graduate Ecology Program at Colorado State University invited H.T. to present a general lecture and graduate seminars. He opened his attaché case with over 500 acetate sheets, and proceeded to talk about Ecosystems, the Maximum Power Principle and other big ideas. When a stack of at least 200 of them fell to the floor, he said: "No problem, I will just pick up where I left off," and continued through another > 100 visuals. He was truly a polymath, and the world of ecosystem science lost two giants when Gene died in early August 2002 and his younger (by ten years) brother only three weeks later.

In April 1968, the week that Martin Luther King Jr. died, Dick Wiegert and I took our families by car to Miami, and then flew to San Juan, Puerto Rico, for further research at the El Verde field site. We stayed in a condo at Fajardo Beach, and drove up to the field site every day, while the wives and children (son Bill was 10 months old) relaxed by the pool. Dick and I studied the activities of the Nasute termite, *Nasutitermes costalis*, which lived in ground-dwelling nests in the forest. We isolated all life stages (workers, soldiers, the king and queen as well). To do follow-up research back on the UGA campus, we carefully packed two queens in sealed plastic bags, and because we were young and impetuous, we did not try to get USDA quarantine clearance to carry the queens out of Puerto Rico. We had Fran pack the queens in the bottom of a diaper bag, covered them first with clean and dirty diapers on top, and got them through the security check line with no problem. This ruse enabled us to do some bomb calorimetry, which was the first ever for this termite species, and published subsequently in a short note the following year.

Dick Wiegert was very generous in sharing his ideas and concepts. He had taken the Ohio State University short course in Acarology ca. 1964, and was supportive of my pursuing studies of detrital foodwebs in microarthropod communities. He also was eager to see research done on the compartmental analysis of soil respiration, which could be very informative if pursued by simple measurement of litter, roots and soil components separately, and seeing how close their respiration (measured separately) came to that of intact cores. By 1968 and 1969 he was beginning his several years' long study of hot springs foodwebs at Yellowstone National Park. The soil respiration analyses were a natural fit with my field studies at SREL, so I pursued this topic with undergraduate support and technician help across several years at SREL. The principal finding was that one could measure root, soil and litter contributions to total soil respiration, by measuring their carbon dioxide outputs separately after first measuring the soil respiration outputs from intact soil cores (Coleman, 1973 a and b).

At the time of the Soil Ecology meeting at UNESCO in Paris, I presented a paper written by Gene Odum on terrestrial energetics. A few weeks after the meeting, Gene, Dick Wiegert and I met to discuss how best to write up a final version of it, as Gene Odum was too busy to be senior author. This paper made special reference to the concept of soils as an "external rumen," analogizing organic matter decomposition in soil to the action of the rumen of a cow. Dick had not attended the meeting in Paris, but was eager to get the idea and supporting concepts out into the literature, seeing it through to completion (Wiegert et al., 1970). Dick passed away in November 2002, a further loss of a major figure in ecology.

My research on fungal foodwebs in old fields arose from the comparison and contrast with Dick's earlier work with the stem well technique for introducing P-32 into plant tissues (Wiegert and Lindeborg, 1964). In 1970, Dac Crossley and I were determined to find easier, more rapid ways of introducing isotopic 32P into the broom sedge grasses in Field 3-412 on the SRS. We proceeded to apply the isotopic solution using a "flit gun," or insecticide sprayer, that got considerable radioactivity onto the plants, but also, in the very humid midsummer climate, it spread onto our lab coats, the steering wheel of the truck, and parts of us. This was an experience not to be repeated. During my six months of work with Francis Clark in the spring and summer of 1971, I mentioned our results to Alicja Breymeyer, a visiting researcher from Warsaw, Poland, and Drs. Jack Lloyd and Bob Lavigne, entomologists from the University of Wyoming who were part of the IBP Grassland Biome research team. To improve on the SREL results, we proceeded, very laboriously, to paint 32P solution onto literally hundreds of individual grass leaves of Blue grama (*Bouteloua gracilis*), using small artists' paint brushes, and followed the isotope through the blue grama root systems and on into root-feeding arthropods. We counted significant radioactivity in root feeding insects of the family Margarodidae, and wrote the study up in the IBP Technical Report series (Coleman et al., 1973).

22.
Major change in the Coleman family

In the final months of our life in Aiken, SC, we adopted a young Korean-American child. The process of obtaining our son Dave was much more complicated than growing one of our own, as it were. We were put in touch with an adoption agency in Pennsylvania called Welcome House. Among several pictures of young children, we picked out one of a two year and nine months' old child. He was infected with tuberculosis, and being treated with strong antibiotics. It was touch and go whether he would be released to emigrate. Finally, he came with a group of about 30 other young children, arriving in a huge Northwest Orient Boeing 747 flight landing in JFK airport. Uncle Charlie and Fran's mom helped us get him to Philadelphia, and then we flew down to Atlanta and Augusta to get him settled in. The next several years were a real roller-coaster ride of physical health problems and dealing with a smoldering anger he had at being abandoned by his mother and relocated more than half-way around the world. He was like my sister Margery. In my mother's vivid phrase, he is a "hard starter and a slow stopper." In less than six weeks' time, we relocated across country, the 1500 miles to my new job in Fort Collins, at Colorado State University. That was a lot for a young boy and ourselves to deal with. He lives in Athens near us, and has some health issues, but is generally mellow.

23.
A retrospective look at my years at SRFL

The years from 1965 through 1971 were a pivotal time for the laboratory. I arrived there under the aegis of Frank Golley, who espoused an ecosystem approach to ecology, and strongly supported all field efforts in that regard. When Frank went up to Athens in 1967, he turned the lead over to Dr. Robert Beyers, an aquatic ecosystem scientist who had worked with H.T. Odum, at the University of Texas. Bob was well-meaning, but got caught up in local petty intrigues on the Savannah River Plant (a problem which Frank Golley had specifically warned us against). Emphasis gradually shifted to population ecology studies, led by Michael Smith and Whit Gibbons, working on small mammals and reptiles, respectively. We hired Claude Boyd from Auburn University, who worked on aquatic nutrient cycling, but he left a few years later. By late 1972, the year that I joined the Grassland Biome group in Fort Collins, CO, Gene Odum named Mike Smith as the director, and SREL continued on the population trajectory ever afterward.

24.
The formative years of the U.S. International Biological Program (IBP)

*A*nother big name on the ecosystem scene was George M. Van Dyne. George dropped by SREL in 1966, and presented his version of big system simulation modeling to our group. Shortly after that, he moved to Colorado State University and began the US/IBP Grasslands Biome program, initially with seed money from the Ford Foundation and then in early 1967, with funds from the newly established Office of Ecosystem Studies of the National Science Foundation. The program was interpreted broadly to include all aspects of biological productivity in relation to human welfare. Numerous governmental agencies in Europe provided funding for studies that began in the early 1960's. There was considerable interest in the United States for the IBP concept, but no significant funding mechanism existed for it. With the assistance of senior scientists in the biological community, including W. Frank Blair (the University of Texas), George M. Woodwell (Brookhaven National Laboratory) and Arthur D. Hasler (the University of Wisconsin), a series of planning meetings were held during 1966, including a pivotal one in August, in Williamstown, Massachusetts, chaired by Eugene Odum. An action plan was created to establish a series of IBP sites in each of the major biomes of North America, beginning with a Grasslands Biome, followed by several others, including forests, deserts and tundra.

In the final months of the Lyndon Johnson administration, several million dollars were authorized and appropriated by Congress, enabling an Ecosystems Studies program office to be established in the National Science Foundation (NSF). Interestingly, a key player in getting the necessary funds appropriated by Congress was Roger Revelle, who was a non-ecologist (albeit an oceanographer) but had a large influence with members of Congress. As planned, biome research programs were begun, with the Grasslands Biome being established first at Colorado State University (CSU), Fort Collins. This was truly an example of preparation meeting opportunity, because the principal investigator, George M. Van Dyne, was primed and ready for this large program.

Van Dyne grew up on a ranch south of Trinidad, Colorado, almost on the New Mexican border. George, an accomplished horseman who worked on the ranch as a hand, was enamored with all aspects of the West. George earned his B.S. degree in Animal Science at CSU, and then went on for his Master's degree in Range Science at South Dakota State University under Mr. James K. ("Tex") Lewis, undertaking a total system study of rangeland ecology. Van Dyne then received his Ph.D. degree from the University of California at Davis, working with Dr. Harold Heady, developing mathematical models of rangeland systems.

George looked carefully for somewhere to launch his career, and settled on Oak Ridge National Laboratory (ORNL), Tennessee, where Stan Auerbach led the Environmental Sciences Division. Jerry Olson and Bernard Patten had already formed a Systems Ecology group there. George joined them in 1963, and the three of them taught the first Systems Ecology course in the USA at the University of Tennessee in Knoxville. At that time, Oak Ridge was one of the few places in the world that had computers capable of solving the complex differential equation and matrix models being formulated by George and his two colleagues. They were the first to use both analog and digital computers to model natural systems.

Because the three scientists worked full-time at Oak Ridge, they drove to Knoxville in a van on Saturdays, taking turns offering one-hour lectures each, with diverse ideas and methods for studying ecological systems. Van Dyne, being junior and serving in a "clean-up" role, would follow Drs. Patten and Olson during the noon hour, writing on the chalk board with his right hand, eating a sandwich with the left, and talking in his soft, but intense tenor voice about many exciting developments in ecosystem modeling (D.A. Crossley, Jr. pers. comm.). Students who took the course were unanimous in their praise of the creativity and the dedication of these young instructors.

George Van Dyne was equally respected in the Environmental Sciences Division for his high research productivity. He suggested to a delighted Stan Auerbach that two scientific papers per person per month be considered the norm for full-time scientists in a research group. George then proceeded to write up to four papers per month in the 18 months he was at Oak Ridge, drawing upon many data sets he had accumulated throughout his Master's and Ph.D. research. Many of his more than 120 refereed scientific publications were written during his Oak Ridge years.

Drs. Olson, Patten, and Van Dyne were instrumental in developing the concept of systems ecology, a quantitative approach for studying and integrating entire ecosystems, including their biotic and abiotic components. Bernie Patten went on to a distinguished career at the University of Georgia where he developed new theories on modeling ecosystem self-organization, nutrient cycling, and energy transformation. Jerry Olson remained at Oak Ridge, and became widely recognized for his pioneering work on global carbon dynamics.

George Van Dyne moved to Colorado State University in the fall of 1967, establishing the Natural Resource Ecology Laboratory (NREL) to pursue the Grassland Biome studies. He began with a secretary and two graduate students (L. J. "Sam" Bledsoe and R. Gerald Wright), along with an initial seed grant from the Ford Foundation until funds from the NSF Ecosystem Studies office began arriving. Van Dyne followed the top-down management scheme agreed to in the 1966 action plan on how to set up a Biome program, although other Biome Programs were more decentralized. He established a central headquarters at the NREL and an intensive study site (the Pawnee Site) near Nunn, Colorado, located on the shortgrass steppe northeast of Fort Collins on the Central Plains

Experimental Range, a research station administered by the USDA Agricultural Research Service.

Van Dyne's grasp of ecosystem science led him to produce and edit a book entitled, *The Ecosystem Concept in Natural Resource Management* (Van Dyne, 1969). The book contained chapters by various authors, all colleagues of George, who had led advances in systems ecology: scientists like Herb Bormann, Chuck Cooper, Gene Likens, Jack Major, Stephen Spurr, and Fred Wagner.

Within two years, George had a burgeoning program in place. He brought Dr. Don Jameson into the program as assistant director over research at the Pawnee Site. Numerous graduate students worked on a diverse and extensive series of studies designed to understand and parameterize various grassland ecosystem processes. These included such subjects as feeding and assimilation studies of animals ranging in size from arthropods to bison, effects of grazing on both above-ground and below-ground productivity, and plant-water-nutrient relationships on both species and entire plant communities. A dozen extensive, or satellite study, sites were established on grasslands from the Osage Tallgrass prairie in Oklahoma on the east, to the annual grassland in California in the Sierra foothills above Fresno. The sites began generating data in 1969 for the modeling effort. Dr. Norman French, formerly of the Nevada Test Site of the USAEC, joined the Grassland Biome to supervise this extensive network of field sites spanning three time zones.

An anecdote about Van Dyne demonstrates his fabled chutzpah. At one point, he proposed to a CSU Vice President that he needed to acquire a building off campus to provide several thousand more square feet of floor space. The official insisted that only a Dean could make such a decision, not a faculty member, influential though George was. George immediately shot back with: "So, make me a Dean!"

All data, including those from the outlying network sites, were sent to the NREL for archiving and analysis. The Biome's statistical design was to collect adequate field data to estimate population means within 20 percent and with 80 percent reliability. Another protocol required all sites to use the same plot size for estimating plant biomass. The NREL provided each site with screening statistics of submitted data, including the sample size necessary to achieve the above-mentioned sampling adequacy.

Dr. George Innis, a mathematical modeler, joined the Grassland Biome in 1970, and shortly thereafter assembled a cadre of postdoctoral fellows from a variety of disciplines. Several of these postdocs, including William Parton, William Hunt, and Robert Woodmansee, remained at CSU and established distinguished careers as senior scientists in the NREL. They continue to participate in large international programs in East Africa, Asia, and South America. Dr. Woodmansee served as the third Director of NREL between 1984 and 1992.

25.
Joining the IBP/Grasslands Biome Program

By late 1971, it was apparent to both Van Dyne and the NSF that the program's rapid growth had reached a point at which not even George could provide all of the scientific direction and leadership required to make the Biome study a success. To coordinate the numerous ongoing data gathering and modeling activities, four more persons were hired to serve as "integrators" of the project. The initial persons hired were: Freeman Smith, abiotic factors; John Marshall, primary producers; Jim Ellis, consumers; and David Coleman, decomposers. Many years later, Francis Clark, one of the senior scientists in the Grassland Program, revealed that I was one of four candidates for the Decomposer Integrator job. Clark strongly endorsed me, and I was hired soon afterward. John soon returned to his home, Australia, and Jai Singh, Banaras Hindu University, Varanasi, India, took his place as primary producer integrator. The integrators' principal job was to conduct and encourage synthesis in the form of internal "gray literature" publications, called Technical Reports, of which over 200 were produced in just four years' time, and also as refereed journal and book articles. While several of the modeling postdocs stayed on with the NREL, only Jim Ellis of the original four integrators remained to lead the Lab into major international research projects in the 1990's and beyond. Dr. Ellis' tragic death in an avalanche while cross-country skiing in the high Colorado mountains in March 2002 cut short his preeminent work on understanding interactions between natural processes and human societies. Jim's ability to conceptualize and synthesize large, complex systems was second to none, with the possible exception of George Van Dyne.

My first contact with the IBP/Grassland Biome happened while attending a U.S./Canada grassland ecology symposium at

Saskatoon, Saskatchewan, in September 1969. Robert Coupland and several colleagues at the University of Saskatchewan were conducting a large-scale Canadian IBP study of grassland ecology on a former cattle ranch, the Matador site in Kyle, southwestern Saskatchewan. George chartered a DC-3 airplane to fly two dozen scientists from around the west from Denver to Saskatoon to participate in the symposium. The ride up there was incredibly rough, and the ashen-faced participants were uncharacteristically quiet on the first day of the meeting. By the time we moved out to the Matador site, interests and volume of discussions had intensified, as had our collective thirst, which was slaked by many cases of good Canadian lager over the course of the three days of presentations and discussions. During these meetings I met Francis E. Clark of the USDA Agricultural Research Laboratory in Fort Collins. I was to have fruitful interactions with him over the next several years.

The Grassland Biome's "crown jewel" was the development of a total system model, called ELM, an acronym for "Ecosystem Level Model." It took George Innis, many postdoctoral fellows and research associates, including Gordon Swartzman, George W. Cole and others working long hours to produce a very detailed model that had 4400 lines of code, 180 state variables and 500 parameters. It required roughly 7 minutes to compile and run a two-year simulation with a two-day time step on a CDC 6400 mainframe computer. Roughly 20 man-years of effort went directly into its development and reporting.

The structure of the model was somewhat elaborate, including some ecosystem components that might, in retrospect, have been omitted. Further, its stated objectives were somewhat vague, and its real objective was to prove that it could be built. In other words, Van Dyne wanted to demonstrate that ecologists knew enough about grassland processes, mathematics, and systems analysis that a mathematical construct that acted like a grassland system could be developed. In his view, such a construct could be used to examine grassland dynamics in place of, or as a complement to, field experimentation. The issue of the feasibility of developing such a model was very much an open question at that time (Coleman et al., 2004b). Keeping in mind the interest of the National Science Foundation in using new technologies, the whole system model was a very strong selling point in getting funding beginning in the late 1960's.

The model itself may not have been much of a success in terms of being used to answer questions about grasslands, but it was a necessary precondition for the development of simulation modeling in ecology. In that sense it was quite successful. Much more useful simulation models quickly followed the development of ELM. Most of the large-scale, long-term, multi-variable questions now being addressed by those interested in the ecology of our planet would be largely unapproachable in the absence of the modeling capabilities developed and proven in the IBP.

The Grassland Biome years were short and intense, lasting from late 1967 until 1974; one year after George Van Dyne stepped down as NREL Director. They were characterized by many planning meetings and extensive travel; in essence, an experiment in "top down" biology. The travel included numerous site visits by those in leadership positions, and trips to Fort Collins and other central locations for synthesis activities by participating scientists. The interactions and synthesis were, perhaps, the Biome program's strongest contribution to science.

Under the leadership of George Van Dyne and other pioneers in systems ecology, the IBP provided the first broad forum in history to integrate the various disciplines in ecology, soil science, climatology, etc. into a comprehensive representation of grassland ecosystems. Synthesis and collaboration extended to numerous international meetings and symposia. For an overview of the IBP Grassland Biome years, see Coleman et al. (2004b), and Coleman (2010).

The Grassland Biome's weak points were the same as noted above! A detailed cost-benefit analysis was never done comprehensively, but an evaluation of the three senior biome programs, including the Grassland biome, published in Science (Mitchell et al., 1976) was generally critical of the approach, claiming that it was not cost-effective and the scientific findings were not very significant. Interestingly, however, a series of papers published on Grassland biome studies one to two years after the Science critique were widely cited by other researchers in ecosystem science (Cole et al., 1977; Hunt, 1977; Reuss and Innis, 1977; Woodmansee, 1978).

Two synthesis volumes were produced in the late 1970's as well. Norman French (1979) edited a volume reviewing the major findings of the U.S./IBP Grassland Biome study. By 1978, a major international compendium on grassland ecology was assembled, based on several synthesis meetings in the early and mid-1970's, and published two years later (Breymeyer and Van Dyne, 1980).

Grassland Biome members attended many national and international meetings. The Matador site meeting of September 1969 led to a life-long collaboration and interactions with Eldor Paul and Francis Clark. The latter proved his mettle in supporting my rather tentative early studies at SREL of the compartmentalization of soil respiration into litter, soil, and root contributions, which I presented as a short paper to the assembled group at the Matador meeting. During question time, Eldor and George Van Dyne attacked some of my methods and inferences drawn from such a simple approach. Feeling quite nonplussed, I saw a large, bald-headed man at the back of the room waving his hand to be recognized. This was Francis Clark, whom Eldor Paul termed "the dean of North American soil microbiologists." Francis suggested that there might be some merit in my experimental approach, and urged the others not to be hasty in condemning the method. Judging from the deference shown him by both Eldor and George, I realized I had a strong ally in my corner.

Later on, toward the end of that memorable meeting out on the black clay soils of the Matador site, I watched Eldor and Francis go out to a low hill on the site in the early morning, and sit and discuss matters for a long time. I found out they were discussing a landmark among synthesis papers, *"The microflora of grassland,"* which was published in Advances in Agronomy (Clark and Paul, 1970). This paper discussed the principal ecosystem processes associated with grasslands, and had considerable influence on my future research.

In November 1970, Francis invited Victor Bartholomew (from North Carolina State University) and me to serve as advisers to him on his studies out at the Pawnee site and to listen to the results of the annual meeting of the IBP Grassland Biome. Bartholomew was one of the pioneers in the use of stable isotopes of nitrogen to follow key nitrogen cycling processes, and I learned a lot from him. Another invitee at the meeting was Dr. Lech Ryszkowski, from the Polish Academy of Sciences, Poznan. He and I developed a life-long friendship and collaborated on several papers and workshops together.

Francis invited me to work with him on measurements of soil respiration at the Pawnee site for six months in the spring and summer of 1971. I brought wife Frances and four year old son Bill, furniture and some supplies for field work in a U-Haul trailer, the ca. 1500 miles from Aiken, S.C., to Fort Collins in April. I rode a bicycle to and from campus, and drove out to the Pawnee site with either Francis, or rode in the big red Chevrolet Suburban that made daily trips to the site. We set up a series of field experiments, with and without irrigation or rainfall, and determined the respiration rates of various strata in the nearly rock-hard soils out there.

Francis thought big and purchased two foot lengths of 8 inch diameter steel well casing to be pounded into the soil in numerous field plots. After he and I tried to drive the casing in with short lengths of steel rail, he observed that "we need someone with a weak mind and a strong back." I told Francis that I failed on both counts, and he hired a strong young undergraduate and got the job done in record time. This study on respiration in various soil layers was quite an education for me. I worked up an extensive analysis of variance of the hundreds of soil respiration measurements I had taken, with the help of our statistician, Marilyn Campion. Francis said that was fine, but he wanted to see the raw data as well. I took over a stack of large 30 column data forms, suitably labeled, but almost impossible to read through. He spent nearly 20 minutes scanning them all, made a few comments, and seemed relieved that all data were present and accounted for. We co-authored a Grasslands Biome Technical Report (Clark and Coleman, 1972) that served us well in some later synthesis activities in my days as a Decomposer Integrator.

Francis Clark was an inspiration to work with, because he had interests other than the science that he pursued, along with the immediate research. After working for several hours in the hot sun, he would say it was time to take a break and look for arrowheads. We rode across the winding dirt roads of the Pawnee Site in his big government-gray ARS Rambler Ambassador, looking for suitable sites on nearby hilltops. He invited me to imagine the Pawnee braves making arrowheads using flints, and speculated on the large herds of bison that must have been there. We saw old buffalo wallows, and the thought experiment was most enjoyable.

26.
International scientific experiences during the IBP

*I*n July 1973 we flew out to San Diego and left the boys with my parents and Auntie Margery and flew from Los Angeles to London, and then directly on to Warsaw where I attended the Grassland Ecology meeting at Dziekanow near Warsaw.

I met researchers from Russia (Roman Zlotin), Czechoslovakia (Blanka Ulehlova), and France (Roger Schaefer). Working with these colleagues, a group of us interested in decomposition studies wrote up a major synthesis paper on Decomposition Processes in Grasslands, which I co-edited with Prof. Albert Sasson, of the University of Rabat, Morocco. The paper was completed in 1976, and published (Coleman and Sasson, 1980) as part of a large international synthesis volume, called *Grasslands, Ecosystems Analysis, and Man* (Breymeyer and Van Dyne, 1980).

After the meeting we met with Lech Ryszkowski in Poznan (Turew) at his laboratory in a 700 yr. old palace. We then visited Wladyslaw Grodzinski in Krakow (Jagiellonian University), where we went on a day trip on the Dunajec River through the Pienniny Mountains in the south of Poland.

We returned home via Vienna by train. This was a harrowing experience with no food service available. The conductor asking menacingly: "Do you have any Polish money?" Our answer was a resolute "NO!" We continued home via Paris, and London on another big Pan Am 747 flight to Los Angeles, then to pick up the boys and return finally to Fort Collins.

One of the more memorable international trips I ever made for the Grassland Biome was to an International Tropical Grassland ecology conference in mid-January 1974 at Banaras Hindu University, Varanasi, India. Our airfares were paid by State Department "counterpart funding" from Public Law 480, established at the end of World War Two. Our tickets were purchased by the Indian government and issued by Pan American Airlines for travel along any route we could justify scientifically. In early January 1974, I flew westward via San Francisco, Tokyo, and onward with stops in Hong Kong and Bangkok, arriving in New Delhi at four in the morning. I had seen crowding and masses of people on dusty streets in northern Mexico, but nothing like the scenes of northern India.

We flew from New Delhi to Lucknow, and then rode a train for most of the afternoon to the eastward, arriving in Varanasi in early evening. During the five days of meetings we also saw classic Indian dancing, took a field trip out onto the Deccan plateau, and generally soaked in the culture. After the meetings, Jai Singh, Freeman Smith and I rode a train from Tundla Junction (near Varanasi) to Agra. After seeing the Taj Mahal, we hired a cab from Agra, and rode for more than six hours the one hundred fifty miles from Agra to New Delhi. After ingesting the dust swirling up from the mostly dirt roads, we could empathize with John Wayne's film, "True Grit."

The remainder of the trip was a study in contrasts, as I flew via Air India from Bombay (now Mumbai), with a fueling stop at Kuwait City (gas flares eerie in the night sky), and onward via Rome. During a two day stopover in Paris, I met with co-authors (Albert Sasson and Roger Schaefer) of the synthesis article, mentioned earlier, on Decomposition processes in Grasslands worldwide (Coleman and Sasson, 1980, in Breymeyer and Van Dyne, 1980).

From Paris, I flew to Stockholm, Sweden, where Thomas Rosswall, a good friend from earlier synthesis meetings in Fort Collins, met and drove me to Uppsala, where I presented a seminar to the Soil Science Department of the Agricultural University on our recent work at Fort Collins. I managed to endure the snow and cold of the ca. 7 hours of daylight in the far northern winter. The marked contrast between the heat and vegetarian cuisine of India and the pea soup and pancakes with Lingonberry jam served at the University Cafeteria, and steak tartare made by the inimitable Marianne Clarholm at her home was one of the indelible memories of that whirlwind three week trip.

One of the hallmarks of IBP-era studies was the extensive amount of synthesis that was carried out within countries and also internationally. This "leveraged" in effect the amount of work being carried out, and helped ensure a wider audience for the research that was conducted. A case in point comes from my experience with Soils and Grassland-related studies. Dr. John Phillipson convened a major synthesis effort in soil ecology in November 1967. John was an Assistant Professor of Zoology in Durham University. He informed colleagues worldwide that the meeting would be held for five days at the UNESCO Building in the 7th Arrondissement in Paris, France.

Gene Odum, Frank Golley and Dick Wiegert were all interested in our having some representation from Georgia at the meeting, so I worked up some of the data from my research in spring and summer of 1967, using $^{65}Zinc$-labelled fungi that I had traced through food webs of fungal-feeding microarthropods in the principal old field at SREL, field 3-412. The study was eventually published in rough form in an IBP publication (Phillipson, 1970), then more extensively analyzed in Oikos (Coleman and McGinnis, 1970). Of much greater import for my personal professional development, I had a chance to meet some of the "greats" in the field, including Eldor Paul, Dennis Parkinson, and Clive Edwards as they summarized their recent studies. Numerous participants came from Asia and Africa as well. I was impressed by the quality of the paper presented by Dr. William Banage from Makerere University, Kampala, Uganda. Unfortunately, his research and his life were cut short by the repressive regime of Idi Amin in the mid-1970's.

Additional syntheses were undertaken by each of the biomes internationally, as well as in the major processes of concern. Thus a volume on Secondary Productivity was published by the Polish Academy of Sciences in 1967 as the result of a meeting held at Jablonna, Poland in 1966. A Decomposition volume had been planned for several years, to be edited by Michael Swift and Bill Heal, but it was postponed indefinitely due to the writing and publishing of the famous book authored by Swift, Heal and J.M. Anderson, called "*Decomposition in Terrestrial Ecosystems*," published by Blackwells in 1979. An offshoot of this effort was carried forward as a special issue of Oikos (Petersen and Luxton, 1982).

This issue contains a wealth of information on life-history traits, rates of secondary production, and decomposition processes summarized worldwide. It is one of the more impressive and useful syntheses to come out of the entire IBP, and was strongly supported by a group of senior scientists interested in decomposition biology, chief among these being Prof. Dennis Parkinson of the University of Calgary.

(Right to left) D.C. Coleman, Dennis Parkinson and Clive Edwards at the International Congress of Soil Zoology in Ceske Budejovice, Czech Republic.

Dennis played a key role in my professional development over the years, kindling an interest in learning more about fungal ecology and decomposition processes during the IBP synthesis meeting in 1967, and subsequently in many IBP grassland biome synthesis meetings, including the one in Saskatoon and Kyle, Saskatchewan in September 1969, and in several meetings of the International Colloquium of Soil Zoology from the 1970's onward to the present. Parkinson invited me to serve as an external examiner for several of his Ph.D. students, including Dan Lousier, David Wardle, Cindy Prescott, Dan Durall, and Mary Ann McLean. Dennis had tried to get me hired at the Kananaskis Research Centre in 1968-69, but the ultimate arbiter of that was Prof. J. B. Cragg, with whom it was difficult to agree on a suitable salary. Dennis Parkinson continued to be a leading light in soil ecology. His work culminated in several month-long workshops on fungal ecology at the University of Kurunegala in Sri Lanka. He passed away in early 2010.

There were numerous memorable meetings within the USA as well. I had heard about the redoubtable Diana Freckman (later

(From left) Walter Whitford, Diana Wall and Amy Treonis. Soil Ecology Society, Palm Springs, CA, 2003.

Wall), with her many studies of the ecology of nematodes in desert soils\. I met her first at a Desert Biome meeting in 1975, arriving at the Salt Lake City airport along with W. Frank Blair, the US Chairman for the IBP. Hundreds of rabidly-cheering young people were yelling as we exited the plane near the front of the line of departing passengers. We waved happily back at them, and they said: "Not you, get out of the way of our wonderful basketball team!" I knew we were in God's country, however, because on the marquee of the big motel we were meeting in was the message: "Welcome, Deseret Biome!" Diana and I developed many collaborative studies and publications over the years.

27.
Post-IBP Research Experiences

At a summer 1972 international meeting on the biology of Soil Fauna in Louvain, Belgium, I met Dr. J. van der Drift, who had been a mentor of Martin Witkamp in the 1950's. Dr. Drift had written numerous interesting articles on the ecology of decomposition. I asked him about his ideas on the roles of protozoa and nematodes in decomposition. He agreed with me that they were virtual unknowns in the decomposition ecology world, and encouraged me to pursue some controlled experiments in that area. I took his comments to heart and resolved to develop some microcosm studies in the laboratory. This led to the next pulse of research activity, which was carried out by the "Belowground Project," one of the first studies to be carried out post-IBP in the NREL. Our group, including Vern Cole of the USDA-ARS Soil Phosphorus laboratory, Bill Hunt of the NREL, Don Klein in the Department of Microbiology, Pat Reid in the Department of Forest Sciences, and several other colleagues at CSU submitted a proposal to the National Science Foundation in the fall of 1974, which was funded early in 1975. This research project was very productive, with graduate students such as Richard Anderson, Ted Elliott, Russ Ingham, Carole Allen-Morley and Tony Trofymow senior authors on or co-authoring literally scores of papers. The soil microcosm studies were set up heterotrophically with glucose or chitin as substrates, or autotrophic-heterotrophic ones, using blue grama grass seedlings as nutrient sources. The trophic interactions in the soil food web facilitated nitrogen and phosphorus mineralization and subsequent uptake by the plants. Two major review papers (Ingham et al., 1985; Hunt et al., 1987) drawing on much of this work have been cited several hundred times (the Ingham et al. more than one thousand times) and are now ISI Citation Classics. An interesting "now it can be told" story concerns getting the project funded initially. Jerry Franklin, who was NSF Ecosystem program director in 1974-75 (on leave from Oregon State University), had a divided panel vote, with some of the panelists urging no funding because we did not have adequate preliminary data to support our proposed experiments. He said he would write a letter for the file and urged us to work extra hard to justify his confidence in us. Our initial funding of $210 thousand per year was enough to support several senior staff a few months per year and also four graduate students and one post-doctoral fellow. That was one of the better decisions Franklin ever made on behalf of soil ecology, and certainly for my career advancement.

PART TWO:
Later Professional Experiences

28.
Post-IBP Travel

*M*y scientific travel reached a peak in the late 1970's. A typical example could well be in the year of 1978. Pat Reid and I read about a scientific meeting arranged by R. Scott-Russell, in Oxford, England, during the spring vacation for the colleges at Oxford University. This seemed to be a good time to take a "road trip" to visit several scientists in the U.K. that we both knew from other meetings, or wanted to meet. The trip was arranged to make driving visits first to colleagues in Sheffield (David Read and students). Read was a pioneer in studying the networking by mycorrhizae between same and other species of plants, using carbon isotopes to measure the magnitude of these processes. This foreshadowed by decades the extensive studies of mycorrhizal networking by Suzanne Simard (2020) made famous by her fascinating book, entitled *"The Mother Tree."*

We then visited Edward Newman and colleagues at the University of Bristol, and drove up to Oxford, where we attended three days of meetings. I presented a poster paper (Elliott et al., 1979) and Pat offered an oral presentation on his work on forest tree ectomycorrhizas. We had chances to visit with Scott-Russell and various luminaries including the famous Albert Rovira, who studied crop root production and turnover, working in CSIRO, South Australia. Scott-Russell had large bottles of Bell's Whisky on hand, and the discussions were generally very lively. After the meeting, Reid and I traveled to the outskirts of London, parked our car and used surface transport to see plays and enjoy the cosmopolitan cuisine of London. One afternoon, we traveled into East London to visit Dr. Mike Swift in Queen Mary College. At that point the east end was very run-down and waiting for the major renovations at the turn of the new century decades later. We then traveled to South Wales to see some of my old haunts out on the Gower Peninsula, staying in a hotel in "the Mumbles," the area of Swansea that I lived in during my postdoctoral year thirteen years earlier. Our final sightseeing trip was down to Devon, to visit Jo Anderson and family in Exeter.

My second trip of the year was to the Ecological Society of America meetings, held in early August 1978, at the University of Georgia, in Athens. I flew from Denver to Atlanta, and took a shuttle bus from Atlanta airport to Athens on the Sunday at the beginning of the meetings. I spent two days on campus, little realizing that I would get a professorship at UGA in just seven years' time. I returned back to Denver and CSU by flying out of Athens on Southern Airways from the Athens airport. There was one other passenger on the flight, Prof. Forrest Stearns of the University of Wisconsin, Rhinelander. The plane was a DC-9, so they lost money on that flight. Soon afterward, the airline changed over to small propeller-driven planes, and now there is no air shuttle service to Atlanta or any other airport from Athens (2022).

The next major trip and meeting was the World Congress of Soil Science in late August 1978, at the University of Alberta, in Edmonton, Alberta, Canada. Several of our group of belowground project colleagues from CSU attended, including our Postdoctoral, Doug Gould, who had gotten his Ph.D. at the University of Alberta. This was my first International Soil Science Society meeting. Attendance at the weeklong meeting was over two thousand, from many countries in Africa, east Asia, Russia and the British Commonwealth. Many of our IBP colleagues were present, so it was like an alumni homecoming. One memory that stands out was the very positive response our papers received on the enhancement of nitrogen and phosphorus cycling in root-rhizosphere soil microcosms. A professor of Soil Chemistry, Dr. Tinsley of the University of Aberdeen, Scotland, suggested to me that the soil science textbooks would have to be rewritten, after seeing our results. That was a very nice thought, but as the history of soil science turned out, such was not to be the case. Our studies have been widely cited, but more locally within the Soil Ecology community.

Two weeks after the meetings in Edmonton, it was time to attend the second International Congress of Ecology, in Jerusalem, Israel, in early September 1978. This was a chance for our Grassland Biome group to present synthesis papers, and to mingle with scientists worldwide. I shared a hotel room with my old friend from Swansea days, Prof. Paul Berthet. We walked several blocks to the venue at the University of Jerusalem. The food, atmosphere, and social interactions were fascinating. Mr. Teddy Kollek, mayor of Jerusalem, welcomed us to the city and gave an excellent ad hoc commentary about the Yad Vashem Museum of the Holocaust, which had to be seen to fully experience the strong feelings and memories of those horrible times from World War Two. I went with Berthet and several of my Polish colleagues on the post-conference excursion by bus to the south of Israel, which we reached in one long afternoon's bus ride. We visited Sde Boqer, the field site of the Desert Research Center. The founder of the Institute, Prof. Michael Evenari, described the site history back to the days of the early Nabataean Kings in the Second Century C.E, who established rain catchments on hillsides to trap water from the infrequent winter rains. I saw and felt what it was like to be in a truly arid environment, and was fascinated by orchards of pistachio trees, and other crops.

We then visited Eilat on the Red Sea, stayed overnight, and came back via the Dead Sea the next day. We had a large Polish contingent on our bus. Professor Kazimir Petrusewicz from the Institute of Ecology, Dziekanow, regaled us with tales of his life in the Polish Merchant Marine, in the early 1930's. He mentioned spending an evening in a tavern on the waterfront of Jeddah, Saudi Arabia, recalling events of his earlier days to sailors there. Speaking mostly in French, the lingua franca of sailors as well as scientists early in the 20th century, he mentioned that he was from Pologne, or Poland. One of the Arab sailors said: "no, you are from

Lechistan." Petrusewicz was delighted by this, as it harkened back to early days in Slavic history, when the legendary three princes who founded their countries in central Europe in ca. the tenth century were as follows: Lech (Poland), Czesch (Czechoslovakia), and Ross (Russia). Apparently all landed in the center of Europe, and Lech went north, Czesch went south, and Ross went east.

This trip was just beginning, as I continued on from Jerusalem after the conclusion of the meeting, and flew to Athens, Greece, for a few days of sightseeing in Athens, and nearby areas. I took a day sightseeing trip to Delphi, where the oracle used to live, and then a day trip by hydrofoil boat from Piraeus, the port of Athens, to the island of Hydra. I had a bread and cheese and wine lunch and enjoyed a brief foray around the island. The next stop on this epic journey was a week-long trip to Russia. I flew by a roundabout route, taking the Yugoslav national airline, JAT, first to Zagreb in Croatia, then onward to Belgrade. We changed planes to fly non-stop from Belgrade to Moscow. Because those were the days of the Soviet Union, I was met at the airport by an Intourist guide, who accompanied me from the airport to downtown, where I stayed in an old mansion, the Hotel National from the czarist times on Red Square. Security at the Sheremetyevo airport was virtually nonexistent and a total contrast with the very tight security at the airport in Tel Aviv, Israel.

In Moscow, I experienced some of the xenophobia or at the very least, caution concerning dealing with visitors from the West. On the first morning of my visit, I phoned to the Institute of Geography of the Soviet Academy of Sciences, asking for my old friend Roman Zlotin, with whom I had worked in IBP meetings during the early part of the 1970's. The secretary sounded very dubious, and said Dr. Zlotin was not there, and was away on travel. This was very unexpected, as I had been invited (by letter) by Zlotin to visit with him that very week. I spent a half-day sightseeing trip around Moscow, arranged by the Intourist guide, and saw the many famous buildings and the scenic views of Moscow from the high bluffs of the Moscow State University. I then took an overnight train, the *Krasnaya Strela*, or Red Arrow, to Leningrad to visit some scientific colleagues there. I visited with Dr. Aristovskaya at the Institute of Microbiology, Russian Academy of Sciences, distributed several of the publications from our Grassland Biome group, and then spent a half-day sightseeing around the Hermitage Museum.

I enjoyed seeing so many paintings by master artists from centuries earlier that I had read about in my History of Art course at Reed College in the late 1950's. The following day, I flew back to Moscow, and prepared for the second major phase of my Russian visit. In mid-afternoon, I again phoned to the Institute of Geography, and this time was put immediately into contact with Roman. I had some very good visits with him and his colleagues in the Institute. The staff congratulated me on my copious consumption of many cups of strong tea, saying that I drank tea "like a Muscovite." Very few of the scientists there were as fluent in English as Zlotin. Prof. Natalie Bazilevich, a world-famous soil and plant scientist, sighed during one long discussion, saying "Tol'ko po Russki" (only in Russian). I managed to use some of my basic Russian learned in two years at the University of Oregon in the early 1960's. My abilities were only equivalent to that of a Russian high schooler. I managed to keep going for a half hour when they graciously relented and conversed with me in English once again.

Roman insisted that I make time to visit with the famous Academician Mercury Ghilarov, head of the laboratory of Vertebrate Morphology and ironically one of the leading experts on the roles of soil invertebrates, particularly earthworms. Roman dropped me off at the big Soviet Academy of Sciences building, and I was cordially welcomed in to Ghilarov's big office. After a leisurely chat, we went with a car and driver to the Academy "canteen" as Ghilarov liked to call it. It was a very well appointed restaurant, serving fine food, including big heaps of caviar, which Ghilarov insisted on ordering for us, probably so he could have an excuse to have some as well. All this food, and fine wine, ended up costing about eight Rubles apiece, which Academician Ghilarov signed for just as he would have in a fancy gentlemen's club in the U.K. or the USA. This was an interesting view into a privileged academic life from yesteryear.

Viewed from a long-term perspective, seeing some of the cultural treasures of Moscow, such as the world-famous Tretyakov Gallery, was time very well spent indeed. The early painted icons of Andrey Rublev were a real wonder. There was a separate room containing various busts and pictures of Lenin and early Soviet personages. My *sotto voce* comment to Dr. Zlotin, that there were some modern "icons" as well, was silently agreed to by him. Roman and the department secretary accompanied me everywhere, but we did our socializing principally in my hotel room. I had gifts for them, but presented them only in my hotel. I had brought along popcorn in its own metal popper, and Roman took it home very proudly. He and his son popped it that night, and found it to be even tastier than I hoped it would be.

Apparently the hassle of dealing with censorship from neighborhood-watch sorts of people precluded Zlotin from inviting and entertaining foreign visitors at their home. We ate often at my hotel, and even that proved difficult at times. Roman and I had gin drinks in my hotel room and then went to the hotel dining room. It was never more than half full, but if we arrived at 7:30 or 8, the maître de would say: "We have no tables available." Roman was very upset by this, particularly when we saw a party of Japanese businessmen push to the front, and hand a $10 bill to the maître de. I had little extra funds, but offered to pay a similar bribe; Roman would not hear of it. After an hour or more of more cocktails and conversation, we again visited the hotel restaurant, and were invariably admitted then. By the end of my visit, I was more than ready to return home.

I took several flights (from Moscow to London to New York to St. Louis and finally to Denver), trying to get back without having any layovers. Viewed across the perspective of forty years, I wonder what drove me to have such long and arduous visits to many different countries and colleagues. It kept me away from my family more than was really good for my relationship with my rapidly growing sons, but that was to be changed somewhat with the eventful ten months the entire family spent in New Zealand one year later.

29.
Other scientific travel experiences in the 1970's

*B*eing very gung-ho in the 1970's, John Lussenhop from the University of Illinois, Chicago, and I jointly arranged a one day symposium on *Detrital Food Webs in Soil* for the Ecological Society of America (ESA) meetings in Stillwater, Oklahoma, in August 1979. One of the attendees, Prof. David Frey of the University of Indiana, commented that the papers about detrital food webs in soils were the high point for him of the entire ESA meeting that year. This was followed up a year later by an even more ambitious day-long symposium at the Ecological Society of America annual meeting in Tucson, Arizona, on *Nematodes in Soil Ecosystems*, arranged by Diana Freckman. The pace of the papers was fast and furious, and the discussions over numerous cool beverages were even more hectic.

Perhaps the most frenetic meeting time I can recall was the International Colloquium of Soil Zoology (ICSZ) at Uppsala, Sweden in mid-June 1976. To get the most out of our travel dollars, we (Bill Hunt, Nancy Stanton from the University of Wyoming and me being the grassland contingent) first visited Lech Ryszkowski at his field station in Turew, western Poland, along with David Reichle and Beverly Ausmus from Oak Ridge Environmental Sciences Division. After many presentations and discussions at Turew, we celebrated and danced the evenings away, including imbibing some concoctions of lab alcohol, cinnamon and sugar "instant liqueur" made by Ryszkowski's lab assistants after the vodka ran out. We were in great shape for Uppsala, Sweden as one can imagine, as the meetings ran for five days at a convention center in town. We either walked or rode the bus the 8 kilometers to the campus housing at Ultuna, with nightly sessions in various people's rooms, taking turns imbibing each other's duty-free liquor, and getting on average no more than two hours of sleep per night.

We usually gathered in Mike Swift's room and his colleague John Perfect regaled us with his imitations of many peoples' accents, including our own, of course. Diana was one of the ringleaders in forming a conga-line of more than 100 participants who, after the Conference banquet, led us the 8 km. back through the streets to our dorm rooms.

In the 1976 ICSZ meeting the idea of microbial-faunal interactions having a significant impact on ecosystem nutrient cycling were just beginning to come to the forefront, and these concepts were hotly debated in sometimes lengthy discussion periods after each paper. After the Uppsala meeting, we proceeded to visit Paris, seeing the sights in the late June heat, escorted by the ever-helpful Patrick Lavelle.

We then visited the famous soil microbiologist Yvon Dommergues at the pedological research institute in Nancy. He was ahead of his time by insisting that his graduate students converse with us only in English, and phrase questions appropriately. This was in marked contrast to the treatment afforded (the then-young) Assistant Professor Patrick Lavelle of the University of Paris at the Uppsala ICSZ meetings by his major Professor, Maxim Lamotte. Patrick mentioned at the beginning of his paper that he would speak in English, so as to ensure that the audience would be large. His professor walked out in protest at such behavior. We applauded Patrick for his strong show of will. We also enjoyed singing along with him as he played many folk songs on his guitar in the evening bull sessions at the University dorm rooms.

30.
In the "Land of the long white cloud," a sabbatical in New Zealand

The only genuine sabbatical leave I ever took was one funded by the New Zealand government from September 1979-July 1980. I had corresponded with both John Stout and Gregor Yeates of the Department of Scientific and Industrial Research (DSIR) Soil Bureau in Lower Hutt. Both had visited us at the NREL in 1978 and early 1979. John urged me to apply for the New Zealand National Research Council Senior Research Fellowship, and, much to my surprise and pleasure, it was awarded. I took Fran and our two young sons, Bill (12) and Dave (10). We rented a small house in Wainuiomata, a suburb near Wellington, New Zealand's capital. Our sojourn was most memorable, living in a country of 2.8 million people, and over 60 million sheep. John Stout and I visited universities and research stations from the bottom of the South Island (Dunedin) to the outskirts of Auckland (Hamilton) on the North Island. I presented research seminars at most of them.

The New Zealand government provided funds for airfare for both John Stout and me to travel in-country, which was most generous, as John was a paraplegic and had difficulty getting around on either bus or train. New Zealand Airlines hoisted John aboard in his wheelchair using forklifts, and gave him the royal treatment. We got to visit sites from which John had received samples over the previous decades and had never seen. Some of the more memorable sites we visited were on the South Island, where Gregor Yeates was following the several year migrations of earthworms across improved pastures (seeded to ryegrass and clover) and unimproved ones (with native vegetation only) and their effects on reducing numbers of nematodes in soil (Yeates, 1981). Prof. Alan Mark, of the Botany Department, University of Dunedin took us on a field trip, up in the high grassland country of the Remarkables, a range not far from Queenstown. We saw the bunchgrass, *Festuca novaezealandiae* and other k-selected species, adapted to growing under nutrient limited conditions (Chapin, 1980; Chapin et al., 1990).

While on that trip in March 1980, I crossed paths with my old friend and colleague, Terry Chapin, from the University of Alaska, Fairbanks, who was working on a sabbatical year with Kevin O'Connor, a noted soil scientist from the University of Christchurch. O'Connor mentioned that just before my arrival in New Zealand, H.T. Odum had made a visit to both North and South Islands in a ten day swing around. Odum insisted to O'Connor that soils continued to improve over time, reaching an ideal end point, sort of like the old climax concept in plant ecology. Of course that is not true, as demonstrated by Walker and Syers in a classic paper (1976), where in a chronosequence of soils from 1- 40 thousand years of age, a large proportion of phosphorus was made unavailable (occluded) in iron and aluminum oxides in the older soils.

I wrote few publications during those whirlwind 10 months in New Zealand, but the perspectives gained from meeting and interacting with a wide range of Maori and pakeha (white-skinned) residents of that beautiful country made an indelible impression and shaped the more field-oriented research I pursued for the next quarter century. I also interacted with many physicists and chemists at the Soil Bureau and in the Institute of Nuclear Sciences, which gave me a broader perspective. It was most fortunate that I took the sabbatical then, because John Stout was to live only two years longer. It would have been a great loss not to have spent those engrossing times with both him and Gregor. Dr. Yeates went on to considerable fame as a noted soil ecologist in the antipodes and worldwide. I was most pleased to be able to obtain funds to bring Gregor to the USA to present the keynote address at the 8th international Soil Ecology Society meeting at Callaway Gardens, Georgia, in May 2001. Gregor died at home of cancer in August

2012, at the age of 68.

In Wainuiomata we learned about Maori culture by living next door to Mabel Salmon and her partner Chris Kingston. Mabel is a Maori woman descended from a Maori chieftain grandfather. She delighted in telling us that her grandfather made a practice of eating his enemies when he conquered them. Mabel was derisive when Gregor's mother, one of the elite of Palmerston North, and a descendant of the original founding colony of English settlers in New Plymouth in 1841, offered to sponsor her as a member of the "Group of 1841." Mabel scoffed at that, noting that the Maori were there 800 years before! Of course, the Maori had in their turn displaced earlier indigenous peoples, but she made her case forcefully.

We departed from New Zealand in mid-winter, early July 1980, with some of the mildest weather of our entire ten months there. Truly the blustery, windy weather of the Cook Straits, at the bottom of the North Island, was some of the worst we have ever encountered. Rain typically came at you horizontally, and umbrellas were next to useless in that kind of weather.

31.
Return to Colorado State University in summer 1980

Charles Ralph, Chairman of the Department of Zoology and Entomology (Z and E) at Colorado State University, welcomed me back to a partial position in the department, with three months per academic year covered by teaching funds. He had offered me that position in late summer 1979, and was amazed that I turned it down for the New Zealand sabbatical. For the three months of salary in that first academic year of 1980-81, I taught the introductory ecology course in the Z and E department in the fall while Dick Tracy was on sabbatical at the University of Washington, and a graduate course in Ecological Energetics in the spring. In other years, I also taught the Ecology section (6 weeks) of the introductory biology course in a lecture hall with ca. 350 students. That was considerable work for one-third of a faculty position. Over the next five years, the coverage rose gradually to five months per academic year.

32.
Research and teaching and travel experiences at NREL post-sabbatical (1980-85)

*V*ern Cole of the USDA Agricultural Research Service in Fort Collins began a new initiative in agricultural ecosystems ecology in late 1980. He invited me to join him and several others (Robert Heil, Dwayne Westfall, both of CSU Agronomy Department, and collaborators from the University of Nebraska (Gary Peterson) and the ARS laboratory in Lincoln, NE (John Doran) in a multi-year interdisciplinary study concerning the roles of tillage management in soil micro- and macro-aggregate formation. NSF funded this impressive effort, suitably named: "The Great Plains Project." It included collaboration with John Stewart and Darwin Anderson of the Institute of Pedology, University of Saskatchewan, so it was international in scope.

One of the key sparkplugs in the study was Ted Elliott, who began on a postdoctoral fellowship with this project after having gotten a M.S. degree with Vern Cole in Agronomy from CSU in 1978, and a Ph.D. with Dick Wiegert at UGA in 1982. Ted had begun his scientific career with me as an undergraduate hourly research student in 1975-76 on the end of the IBP grassland biome funding, moving onto the Belowground Project later on. His first scientific paper, based on study of the activity of soil protozoa in the shortgrass prairie, was published soon afterward (Elliott and Coleman, 1977). Ted had that extra spark of curiosity and drive that made him a standout early on. We all enjoyed watching his scientific development take place. The synergism that Ted had with the other students in our project, including Barbara Fairbanks and Larry Woods in Agronomy, and Rick Anderson and Russ Ingham in Zoology, was most productive as well.

Some of the more noteworthy papers Ted wrote include: Elliott et al. (1979), Elliott et al. (1980), and Elliott and Coleman (1988). Ted went on to advise several graduate students, most notably being Johan Six, who published several foundational papers on the dynamics of soil micro- and macro-aggregates in agroecosystems (Six et al., 2000; Six et al., 2004). Johan now has a senior professorship at the Eidgenoessische Technische Hochschule (Swiss Technical University) in Zurich, Switzerland. Ted died of cancer in 2002.

In the late 70's, Mel Dyer, the Ecosystems program head in NSF, said our Belowground project was the third most productive of refereed publications funded by NSF environmental programs nationally. Mel talked on the telephone to Charles Ralph, the skeptical head of the Zoology and Entomology (Z and E) department when I was being considered for a partial academic appointment in 1978. Apparently the shadow of George Van Dyne, who had been long away from heading the NREL, still caused problems with some of the more recalcitrant members of the department. Years afterward, in the early 1980's when I came up for promotion to Associate Professor with tenure, a battle erupted in the faculty meeting considering candidates for promotion. Some of the full professors protested that they didn't want Van Dyne's influence to penetrate the department. Dr. Ralph noted that Coleman's record should stand on its own, but that carried little weight. According to Richard Tracy, one of the junior Associate Professors sympathetic to my cause, it impelled Professor Howard Evans, renowned entomologist and the only National Academy member at CSU, to observe that: "I have listened to this argument long enough. Dr. Coleman has an impressive list of refereed publications and funded research. I vote aye and will not waste my time further." Somewhat surprisingly, the opposition faded, and a majority voted in favor. I had talked to Howard only a few times before; he was a man of few words, but ones well chosen. Upon my return from the New Zealand sabbatical, I asked his advice about teaching the Principles of Ecology course while Dick Tracy was on sabbatical. He correctly noted: "It will be a lot of work; a great deal of work." I enjoyed teaching that course in the fall of 1980 very much. As a further aside about this byzantine series of events, Dr. Ralph confided in me in the early 1980's that the co-director of Z and E, Prof. Wayne Brewer, was foremost among those saying that I was not ready for tenure, as I was an unknown quantity.

Dr. Brewer took a sabbatical in Central Europe, located at Charles University in Prague, Czechoslovakia in 1979. He visited in Brno and at universities in several other cities. He was flabbergasted to hear some of the Czech scientists say they did not know any of his faculty colleagues, but did he know of the work of David Coleman in the NREL at CSU? He came back several months later and said to Dr. Ralph: "I give up. Coleman is obviously known internationally, and I now vote to include him on our faculty," or words to that effect. Perhaps these sorts of interactions are common in deliberations among faculty, but I have seldom seen such maneuvering in my days at the University of Georgia. My conclusion from such an experience is that the "founder effect" of George Van Dyne had quite a carry-over in the general distrust if not outright jealous attitude toward the NREL. Perhaps others in the academic community will see some useful take-home lessons from these interactions. Life was certainly interesting in those hectic years. I say this with the old Chinese curse in mind: "May you live in interesting times."

33.
Speech Difficulties in my Life

With President Biden in the White House, with his self-admitted problems with speech hesitancy, it is timely to discuss my life-long speech problems. Speech therapists, several of whom I consulted with, insisted it was a "stammer, and not a stutter." Whatever the term, it was ever-present in my life, especially in the long years of puberty, when I was expected to be articulate and fast on my feet in classes and also going on the occasional (infrequent) date. I decided to tackle the problem directly only during the latter half of my years at Colorado State University. I worked with Dr. Howard Larrimore, a counselor from the Speech Department, and learned some tricks, such as exhaling through consonants, to get past certain words , and also tackling difficult situations, such as talking on the telephone, which we addressed by my phoning dozens of businesses and asking about business hours, etc. This proved to be a major benefit for my future career, when I went on job interviews and gave seminars, and similar high stress situations.

What marked a turning point for me was the realization that trying to ascertain why the speech difficulties happened was beside the point; how to speak fluently with whatever helping methods was much more fruitful. My dear wife and both sons were very supportive and understanding, which helped facilitate the process as well. Being a life-long academic, who taught classes and seminars for decades, there was no escape from working around this frustrating difficulty. By the time of my retirement in late 2005, I apparently had no need to be defensive about my academic career with over 300 refereed publications. The problem just faded away.

As a footnote to the political maneuverings in my last year at CSU, the NREL was trying to get a new Director appointed by 1984. Both Bob Woodmansee and I tossed our hats in the ring, as it were, and had interviews with Jay Hughes, the Dean of the School of Forest Resources. Unbeknownst to Hughes, we had an election in the NREL, and I came in one vote short in the balloting. Upon discussing this situation with Charles Ralph in Zoology, he expressed surprise that I was a candidate, as Dr. Hughes told him that the one candidate being considered was Bob Woodmansee. What we in the NREL had not realized was that this was to be a "Head" position, not a chairman, elected by the faculty. Hughes appointed Bob Woodmansee, and I once again dodged the bullet for taking an administrative position. Soon afterward, I interviewed at the U of Georgia and my career took a different trajectory.

34.
Seminars and interview for Professor position at the Institute of Ecology, University of Georgia

In the summer of 1981 at the Ecological Society of America meetings in Bloomington, Indiana, I was approached by both Dick Wiegert and Larry Pomeroy about considering applying for a full professor position at the University of Georgia (UGA) in the Institute of Ecology. The position would become available upon the retirement of Gene Odum in early 1984. I could scarcely believe I had a chance at such a great position, but applied for it when the advertisement appeared in Science. I coauthored a major review article with Pat Reid and Vern Cole on "Strategies of Nutrient Cycling in Soil Systems" (Coleman et al., 1983) to develop my curriculum vitae a bit more. The advertisement for the UGA position appeared in Science journal in the fall of 1983, and I applied for and made the short list of candidates, including Paul Ehrlich of Stanford University, and Denis Owen of the Oxford Polytechnic Institute. I gave my seminar at the Institute of Ecology in early March 1984 as a trial run of a paper I had been invited to present to a special meeting of the British Ecological Society in April, entitled: "Through a Ped darkly, an ecological analysis of root, soil, microbial and faunal interactions" (Coleman, 1985). The seminar seemed to be received well by the eager young faculty at the Institute. My name rose to the top after some difficulties were encountered in meeting what Paul Ehrlich (the first choice) and his wife required as a combined salary package.

In early April Fran, Diana Freckman and I drove up to York for the British Ecological Society (BES) Plant and Soil interactions meeting. We stayed at the Swan Inn (made famous by William Wordsworth) in Grasmere en route. I gave the keynote address to that meeting. We drove back with Diana to London and Fran and

I continued on to Switzerland to meet with Marcus Bieri at the Eidgenoessische Technische Hochschule (ETH) in Zurich. We had an epic weeklong trip around Switzerland by Swissrail pass, seeing marvelous glaciers, lovely valleys around Lucerne and Berne. The fabled Swiss Railways lived up to their reputation. Always punctual, never an "Alll aboooard," simply departing on time to the second.

Dean Jack Payne, Dean of the Franklin College of Arts and Sciences, had been out of town when I interviewed in March, so I flew back to visit with him in late summer 1984. He scanned my C.V., and delightedly exclaimed, "Why, you're a microbial ecologist! Of course we want to get you onto our faculty." He was as good as his word, writing up a very detailed three page offer that CSU couldn't begin to touch, including funds for a technician and a secretary. I accepted in October of 1984, and the last twenty years of my scientific career began to unfold as a full Professor.

A brief aside is appropriate here, to recount the history of what came to be the 1985 review paper, "Through a Ped darkly…." At a meeting on "Invertebrate-Microbial Interactions," organized by Jo Anderson at the University of Exeter in September 1982, David Wynne-Williams of the British Antarctic Survey and other eager young terrestrial ecologists approached me to present a keynote paper at the British Ecological Society meeting planned for Plant-microbial-faunal interactions at the University of York in April of 1984. I racked my brain to come up with a suitable name for such a paper, and settled upon the "Through a Ped darkly" theme, drawing shamelessly on the imagery of St. Paul in his First Epistle to the Corinthians, in which he described the works of God as being viewed "through a glass darkly."

I happened upon the quotation from Wallace Stevens' collected poems, in "Man with the blue guitar," and figured I had the metaphorical side of things sewn up. The quotation is:

"Throw away the lights, the definitions,
And say of what you see in the dark

That it is this, or that it is that,
But do not use the rotted names.
How should you walk in that space and know
Nothing of the madness of space,
Nothing of its jocular procreations?
Throw the lights away. Nothing must stand
Between you and the shapes you take
When the crust of shape has been destroyed."

At the end of my hour-long lecture, the floor was open for questions. There was an awkward pause, and then one of the senior scientists from the British "Establishment," Dr. Bernard Tinker, fired a question at me concerning a comment that I had made about a complex, but realistic diagram of Bill Hunt et al. (1987), in a model of nitrogen flows in a shortgrass prairie detrital food web being "God-awful complicated.," to which he appended: " so why even bother to model it?" or words to that effect. This occasioned an audible intake of breath by the assembled crowd of more than 150 persons, who waited for my response. I was delighted to say: "because the trophic interactions really do matter," and was pleased that the audience agreed with me. Top marks to Bernard for asking; many soil chemists and not a few physicists have been quite dismissive of studying this level of complexity in soil systems. I would like to think that perhaps one of the contributions our groups in Colorado and Georgia have made is that the biological interactions in soils have a significant impact on soil biogeochemistry. The new generation of soil ecologists is demonstrating ever more cleverly and elegantly further details of these processes (Coleman, et al., 2018).

In the midst of scientific sessions, the meeting at York was also memorable because of the cultural and historic aspects to be explored. The National Railway Museum is at York, so of course Fran and I had to feast our eyes on the many famous locomotives and rolling stock. In addition there was a fascinating tour of the "undercroft," the lower layers of the York Minster, which exposed, from bottom up, artifacts dating to Roman occupation, followed by Saxon and then medieval times.

Looking back across many years of my professional life, there has always been a dichotomy between the population and evolutionary biologists in ecology and the energy-flow and nutrient cycling researchers. The separation exists today, but seems to be a more tolerant one. Perhaps it is due to the efforts of many of my generation pointing out the many linkages between the two approaches. For example, in the early days of our "Belowground Project," which ran from 1975-1985 at Colorado State University, we measured and observed numerous instances of the impact of population and community-level phenomena, i.e., who eats whom, and the consequences of this activity, on enhanced plant growth and nutrient (nitrogen and phosphorus) content of grassland plants. Much of this activity was localized in the ca. 1-2 mm. zone of soil surrounding the plant roots, the rhizosphere (Ingham et al., 1985).

Several international programs arose after the IBP; some of them, including the International Geosphere Biosphere Program, operated for decades. For instance, in September 1984, Dr. Lech Ryszkowski and his colleagues in the Polish Academy of Sciences convened an international workshop on Soil Biological and Chemical Processes in Managed Ecosystems. The meeting was held at a country inn (the "Duck") near Turew, about 40 km outside of Poznan in western Poland. The main topics of discussion concerned the linking of aboveground and belowground productivity, and the impacts of root-related processes on secondary productivity and nutrient cycling.

Soon after attending this meeting, I visited with colleagues from the Institute of Ecology in Warsaw. Alicja Breymeyer, a long-term colleague and friend, took me on a sightseeing trip in downtown Warsaw. We climbed up in a church steeple to look out over the city. It was approaching dusk, and much to our surprise, we saw literally tens of thousands of people carrying candles, marching on the downtown region. The experience was visually remarkable and aurally so as well: the marchers were completely silent. They were marching in remembrance of a priest, Father Popiełuszko, who had been gunned down by security forces a week earlier. The collective concern and outrage was visually dramatic, and Dr. Breymeyer urged me to walk with her very rapidly to a friend's apartment nearby so we could be away from any harm, if the security people became trigger-happy. Fortunately, there were no incidents, and the demonstration faded away a few hours later.

I continued on my trip to visit other colleagues. As was typical for an invited scientific participant, I had no problems dealing with the Polish passport authorities. En route home, I flew on Interflug, the East German airline, to East Berlin to visit Dr. Ekkehard von Törne, Editor-in-Chief of the journal *Pedobiologia* (I was on the Board of Editors at the time). As the only "Westerner" of the approximately 60 passengers to disembark from the plane, I went through a separate passport line, and was greeted by Dr. Törne, and had a pleasant and informative stay in East Berlin. In the course of the visit, I met Professor Dr. Bernhard Ferchland, the publisher of *Pedobiologia*, and saw the sights of East Berlin, including the famous Berlin Wall.

Getting out of East Berlin, however, proved to be much more difficult than gaining entry. I walked from my hotel to a station on the Stadtbahn (Berlin's light-rail system), or "S Bahn" as it is called. Hoping to catch the next S Bahn to West Berlin, which ran every 20 minutes, I began running along the main concourse, and was promptly intercepted by a tall security official wearing a white coat, who shouted: "Halt, mein Herr!" I did as he instructed, noticing that many people were casting pitying glances my way as they strode past. After establishing that I was an American, the security guard asked me why I was there, what I had in my briefcase, and what possible business could I have that would require me to run through the station at 6:15 in the morning? I tried to respond carefully, since I was carrying in my coat pocket a personal letter from my Polish colleague Ladd Grodzinski that I was to mail to his sister in Massachusetts once I returned home, and I didn't want it to fall into the hands of the authorities.

I proudly (and naïvely) showed the security official a reprint of a paper from an East German agricultural journal, to demonstrate my interest in international scientific collaboration. His response was to turn red in the face and, with spittle flying, yell at me, "Have you any more East German documents?" The fat was truly in the fire by then, and I began to wonder how long I might be detained under less than ideal circumstances. He then inquired if I could prove that I was there merely to visit scientists and had no other ulterior motive. Professor Ferchland had given me a business card the day before, so I presented that to the official. He said "we'll see about that," and placed a phone call to the Professor. It was gratifying to see that the ensuing conversation was very one-sided, with the security man holding the receiver farther and farther from his ear as he received a severe dressing down for inconveniencing a scientific colleague. I was then whisked through security, and made it to West Berlin, and subsequently back home, safely. Those fifteen minutes spent in "limbo," as it were, were etched indelibly on my mind.

As a coda to this story, a few days after my return to the campus of Colorado State University, an official in Fort Collins telephoned me, saying he needed to discuss some important matters. I invited him to my office and thought little of it, until he arrived. He said he was with the Central Intelligence Agency, and asked if I had traveled to Poland recently. I said I had, explaining that I had attended scientific meetings there several times in the previous decade. Among the other questions he asked were: "Are there many horses in western Poland? Did you see many tractors?" He apparently wasn't amused when I responded that I had seen some horse-drawn carts and observed some Poles with a few racehorses. This conversation was more amicable than the one in East Germany, but hardly more productive. The Cold War is long gone, but even in current times there are numerous hurdles for scientists to clear in order to engage in collaborative research. Colleagues in Australia's Commonwealth Scientific and Industrial Research Organisation occasionally have experienced considerable difficulty in gaining access to the US to attend scientific meetings as invited participants.

Here at the University of Georgia, an Indian postdoctoral fellow – after having returned home for a two-week visit – was denied re-entry into the US for over 2 months due to pending administrative clearance regarding a visa. In 2004, during a visit to colleagues at the University of Kunming in Yunnan, China, to see the famous Xishuangbanna Tropical Botanical Garden, I offered to provide them space in my laboratory if they chose to visit the US. After an awkward pause, the scientist to whom I made the offer said, in so many words: "Thank you for the invitation, but getting approval and a visa from the US Embassy in Beijing to travel to your country would be the greatest obstacle to my taking you up on your offer."

Other scientists, particularly those from Middle Eastern countries, have expressed similar sentiments. The regulations of the US Department of Homeland Security and the State Department are understandably strict, but the ways in which they are administered are clearly in need of further refinement. This raises the more general problem of how censorship can hamper international cooperation in science or any other academic endeavor.

From personal experience gained decades ago, while working at a research site operated by a contractor for the now defunct US Atomic Energy Commission, I observed that the profiling of "suspect" groups and even whole countries inevitably leads to missed opportunities and excessive red tape. The entire senior scientific staff at the Savannah River Ecology Laboratory was "invited" to watch a security alert film, which warned us that any contact with persons behind the Iron Curtain was unwise. Upon asking the Commission's security personnel if a recent reprint – sent to me by a senior scientist in Soviet Russia, and which had the cryptic comment "I will send the second reprint later" penciled across the top – prompted the security alert, he said yes, they were aware of that message. Perhaps the rise of social media and opportunities to communicate directly via online technologies such as Skype will facilitate the sorts of international collaboration that is so much desired now. A more general effect of the tightening of international borders is the adverse impact on economic efficiency, and the ability to deal with global environmental problems and numerous other issues that we face in the 21st century. Despite these restrictions, the increase in global trade has facilitated the introduction of invasive plant and animal species that has become increasingly costly to mitigate over time. I hope that, if nothing else, these experiences will stimulate ecologists to enter the fray as we confront ever more daunting challenges in a world that has finite limits and requires collaborative, internationally based approaches to solve its problems.

35.
Years at the Institute of Ecology, University of Georgia

When I arrived in Athens in July 1985, the Institute of Ecology operated much as it had since its inception: a venue where colleagues could interact on various research projects that were both regional and national in scope. Most of the faculty had various National Science Foundation grants through either Ecology or Ecosystems program offices, and some had funding from Department of Energy (DOE) either via the SREL or national programs. Paul Hendrix, Dac Crossley and I were eager to pursue detrital food web studies in agricultural soils (Hendrix et al, 1986). We submitted a proposal to the NSF and were funded for three years to study the differing effects of fungal vs. bacterial-dominated food webs. Elaine Ingham joined us as a postdoctoral in the first year, and Weixin Cheng and Mike Beare began their predoctoral work as well, with the former working on root production and turnover, and the latter on litter decomposition processes at Horseshoe Bend. In addition, I applied to the Ecology program of NSF for funding on further work on detrital food webs in soil microcosms, with Dave Wright as postdoctoral on that study. All this, plus occasional trips to Washington to serve on a National Research Council Committee on *Alternative Agriculture* (I was nominated by Gene Odum, who kindly referred me to them) kept me very busy. This synthesis effort resulted in an influential volume, entitled: "Alternative Agriculture," (Pesek et al.1989).

In my last year at the NREL, I had discussed some initiatives on the study of conjoint cycles of carbon (C), nitrogen (N), phosphorus (P) and sulfur (S) in ecosystems with Ted Elliott, which we were able to interest Jim Gosz at NSF to fund. We held a four-day workshop in April 1966 on Biogeochemical cycling of C, N, S, and

P in the old manor house on Sapelo Island, Georgia. Gene Odum gave a final wrap-up talk, and the interactions by the 35 or so invited speakers who came from more than one dozen countries were very positive.

With scientists and the general public totally accustomed to the instantaneous response rates possible using e-mail and the worldwide web now, back then we had to contact people by air mail or by Telex, as international telephone calls back in 1985 were too costly for routine usage. Many times either Elaine Ingham or I went over to the Science Library to use their Telex machine to contact scientists from more than twenty countries whom we had invited to this workshop. The workshop papers were published in a special issue of *Biogeochemistry,* volume 5 in 1988. One of the key participants was Malcolm Oades, whose work on mechanisms of soil micro-aggregate formation was impressing scientists worldwide. I had further positive interactions with him two years later (see below).

During the first year of my tenure at UGA as a senior scientist, I was asked by Frank Golley to present the first lecture in what has become an annual event, the Odum lecture. The lectures have reached 35 in number, and feature ecologists invited from around the world.

Another collaborative effort that bore considerable fruit in the late 1980's was a project that Diana Freckman (now Wall), Mel Dyer, Sam McNaughton and I discussed with NSF officials Patrick

Fran and Dave Coleman, November 1985 after Dave gave the first Odum Lecture at UGA's Institute of Ecology. (Photo courtesy UGA Odum School of Ecology)

Flanagan and Jim Gosz at the Ecological Society of America meetings in Minneapolis in June of 1985. Out of that arose a three-year project (1986-89) to follow the movement of new photosynthate in "real time" in shoots and roots of African grassland plants (*Panicum* spp.) that Sam had available from his greenhouse at Syracuse University. The main "hook" or scientific point of interest was Carbon-11 generated by a Van de Graaf accelerator at Duke University. The labeled carbon dioxide was transferred by photosynthetic fixation to plants growing under controlled temperature and light at the Phytotron at Duke. Dr. Boyd Strain, our scientific contact at the Duke Phytotron, facilitated our work as we dealt with experimental regimes that required obtaining batches of high-radioactivity Carbon-11 labeled carbon dioxide for labeling our plants. We ran complete experiments in five or six hours, following the movement of the gamma-emitting Carbon-11 as it was fixed, transferred through the leaves and into the roots ("translocated") and ultimately into mycorrhizal tissues. We demonstrated that mycorrhizae were significant carbon sinks, with the enhanced sequestration occurring within a few minutes' time.

Unfortunately, Carbon-11 has a 20.3-minute isotopic half-life (time to decay to half its original value), so the work had to be carried out both carefully and rapidly. We got some interesting papers from this work on mycorrhiza as carbon sinks (Wang et al., 1989; Dyer et al., 1991), but it was logistically too demanding to appeal to a wide range of ecologists. Carbon-14, which has a very long half-life, of ca. 4,800 years, emits beta radiation, requiring that the tissues of interest be processed in some way to measure the radioactivity in a scintillation counter. One might think that the more readily detectable gamma radiation from the Carbon-11 would make it more popular as a radiotracer of interest. Unfortunately, its very short half-life makes it only marginally useful for most purposes. Perhaps more significantly, the revolution in measurement techniques, including being possible to measure Carbon-13, the heavy isotope of Carbon, in real time, makes it the preferred isotope for most metabolic and similar measurements in biological and ecological investigations.

Some of our experimental results served to stimulate experiments carried out by Weixin Cheng on the Horseshoe Bend project, in which he studied the effect of living roots on organic matter decomposition. The higher microbial activity stimulated by the root rhizosphere in turn stimulated labile organic matter decomposition (Cheng and Coleman, 1990). This was followed up with additional NSF funded research on detailed analyses of root and microbial respiration in the rhizosphere, using membrane columns and cells to allow a partitioning of the sources of respiration (Cheng et al., 1996).

I was pre-conditioned to the usage of short half-lived isotopes, because one of the papers John Stout and I published from my sabbatical year in New Zealand reported on our studies of gaseous evolution from cores of pastureland soils in New Zealand, using Nitrogen-13, with a 10 minute half-life (Stout et al., 1984). We used a Van de Graaf generator to produce the N-13, courtesy of the Institute of Nuclear Sciences in Lower Hutt. The technique enabled us to discriminate between labeled ammonium and nitrate and follow its evolution as nitrous oxide over a few minutes to hours' time.

With considerable encouragement from Patrick Flanagan, head of the Ecology program of NSF, Ben Bohlool and I in 1987 prepared jointly a proposal for an international workshop on "*The Dynamics of Soil Organic Matter in Tropical Ecosystems*," which was held in Kahului, Maui, Hawaii in October 1988. We invited a star-studded cast of scientists from Europe, North America, India, Japan, Australia and New Zealand. Price of attendance (including a plane ticket) was a rough draft manuscript that could be assembled into co-authored chapters related to the physics, chemistry, and biology of soil organic matter formation. Obtaining support from the NSF was difficult, despite Dr. Flanagan's strongest efforts. This was because Hawaii and boondoggle seemed to be considered synonymous in the administrators' minds. When Bohlool and I pointed out to them that meeting in the mid-Pacific was most economical for travel by our Asian and Oceania participants, and that people would be accommodated in double rooms, at $39.95 per room, breakfast included, they relented and allowed the planning to proceed.

The output of the workshop was one of the more useful synthesis books I have ever had the pleasure of editing, with Drs. Malcolm Oades and Goro Uehara as co-editors, published by the University of Hawaii Press (Coleman et al., 1989). Being on beautiful Maui, we set up our program to meet in mornings and evenings, with afternoons for walks or surfing. Patrick Lavelle (then at the University of Paris V) brought his guitar and the late evening sing-alongs were also most pleasant and memorable.

In the spring of 1987, John Hobbie and Brian Fry sponsored a workshop on Stable Isotope methods in Ecology, at the Ecosystem Center in Woods Hole, Massachusetts. Jerry Melillo and Eldor Paul were arranging for a series of books to be written on the use of isotopes in ecological studies, and they invited Brian and me to edit a book on all aspects of carbon isotopes. This resulted in the publication of a reasonably successful book, entitled: *Carbon Isotope Techniques* (Coleman and Fry, 1991).

36.
Repercussions from leaving CSU

*I*n August 1988, several of us in our soil ecology group in Athens flew out to Sacramento for the Ecological Society of America annual meeting at University of California, Davis. We flew from Atlanta to Denver, changing planes continue on to Sacramento. During the wait between flights, I picked up a copy of the Sunday *Denver Post*, with a headline story about retention problems of senior faculty at Colorado State University. The article featured prominently the recent departures of Pat Reid, one of our belowground project group in 1986, and myself, in July of 1985. The article asked what was wrong with CSU's retention policy, noting the numerous grants that both Dr. Reid and I had over our several years at that institution. This was too late for any corrective action, but was balm to my ego from the long struggle to get a bit more than half of an academic year's support from that benighted administration. When I interviewed at UGA in March 1984, Bill Wiebe, one of the founding members of the Institute of Ecology, assured me that they would treat me right, when I noted that I still had only five months of support during the academic year.

37.
Early Experiences in Tropical Ecology: Studies at Tsavo National Park, Kenya

One of my early forays into tropical ecology was a memorable two-week trip to East Africa in November of 1986, again with the support of Patrick Flanagan in the National Science Foundation. He was considering funding a research grant by a group from Cornell University led by Drs. Joy Belsky and Susan Riha. Their objective was to account for the causes and consequences of the rings of greener grass that appeared under the canopies of acacia (*Acacia tortilis*) and baobab (*Adansonia digitata*) trees in the East African savanna. Flanagan's precondition for this work (largely soil chemistry and plant community analysis) to be funded was for them to include me, at no extra cost, as member of a sort of "SWAT team" to study soil microbes and nematodes in their study sites. We flew by Pan Am across Africa to Nairobi (with intermediate stops in Dakar, Senegal; Monrovia, Liberia; and Lagos, Nigeria).

After getting the necessary supplies in Nairobi, we traveled south by vans and Land Rovers to Voi, near the entrance to Tsavo National Park. We set up numerous experiments that Sam Mwonga, a very capable and persistent Kenyan graduate student, carried out there over the next two years. This resulted in a short paper describing the nematode communities and their fluctuations under wet and dry conditions, which appeared in *Biology and Fertility of Soils* (Coleman et al., 1991).

The guesthouses in which we stayed had unusual "screens" on the windows, of one-inch mesh wrought iron. This allowed the entry of incredible numbers of beetles. With all the large mammals around there, including buffalo, impala, elephants, etc. generating dung, the land was heaven for scarabaeid beetles. We were instructed to stay indoors during nighttime, due to the presence of marauding lions and other predators. The Cape Buffalo were the greatest menace, due to their bad tempered natures. While doing fieldwork in Tsavo East, we camped out one night near the shores of Lake Jipe, which was home to numerous hippopotami. Even if one had a full bladder, it was best to stay inside the tent. I awoke at 2 a.m. to hear loud snuffling and grazing noises of a hippo outside my tent. When the hippo defecated, it spun its tail around and pelted my tent with the near-liquid dung. That was ample incentive to remain inside the tent!

After our fieldwork in Tsavo East, Susan Riha and I returned from Voi to Nairobi by the overnight passenger train that went from Mombasa on the coast, up to Nairobi. The train was most comfortable, and the meter-gauge train rode well on the roadbed (insert picture). The old mirrors in the railway stations had the caption: "E.A.R," denoting the former East African Railway, which ran from Kenya into Uganda. In the Nairobi yards, I saw retired

Setting up camp at Lake Jipe, Tsavo East National Park, Kenya, Nov. 1986.

Overnight train from Tsavo up to Nairobi, Nov. 1986.

Garratts (articulated steam locomotives in their maroon splendor). Once in Nairobi, we took a trip out to Nairobi Park, just outside the downtown area, teeming with many native mammals (insert picture), and also a short trip by car across the Rift Valley and into the uplands, with their large coffee plantations. We saw many flame trees in bloom, reminiscent of the famous book: *Flame Trees of Thika.*

A second visit to East Africa was made possible fourteen years later by an invitation from Mike Swift to be part of a review panel assessing the research progress of the Tropical Soil Biology and Fertility (TSBF) project. This research group, funded over several decades now, has support from the Rockefeller Brothers Foundation and numerous other private foundations. It is headquartered in Nairobi, Kenya. This project grew out of long-term collaboration between a group of European ecologists, including Swift, Bill Heal, Jo Anderson, and the Swedish microbial ecologist, Thomas Rosswall, among others. Their principal objective was to use principles of soil ecology and nutrient cycling to help increase crop productivity and yield, using intercropping, green manuring, and a

Wildlife at Nairobi National Park, Nov. 1986.

variety of other approaches.

They have an extensive network of collaborators in Kenya, Uganda, Zimbabwe, and several other African and South American countries. Mike Swift was director for over fifteen years, and then a native African was named to succeed him.

During May of 2000, I flew over to join two other reviewers, one from the W.K. Kellogg Foundation, and a recently retired scientist from Consultative Group on International Agricultural Research (CGIAR). We interacted with several collaborators on farms and also discussed research interests in common with colleagues such as Dr. Pedro Sanchez, Director of the research organization, ICRAF, the International Cooperative on Agro-Forestry, also headquartered in Nairobi. We flew to field sites in western Kenya, and also to Uganda during a very busy ten days' consulting. I was captivated all over again by the wide-open nature of the landscape and also its people. To enable us to keep focused, Dr. Swift very kindly flew us by small private plane to Maasai Mara, where we "camped out" in a tent with catered meals and a chance to view large game animals such as giraffes, zebras, hippopotami, etc. The main limiting factor was and still is the lack of access to adequate supplies of nitrogenous and phosphorus fertilizers, a condition that continues to be exacerbated by a very rudimentary road network. Using principles of alternative agriculture, promoting greater recycling of the limited nutrients available in small farm-holders' plots, has proven to be a very effective way of providing a better livelihood for many farmers in nutrient-limited localities.

The security risks during my time in Nairobi were much greater in 2000 than in 1986. *En route* over to Kenya, I changed planes in Amsterdam. Pedro Sanchez, an old friend, hailed me in the airport and we sat near each other in the business class section of the plane. As we were deplaning in Nairobi, he leaned over and said: "Don't go outside your hotel unaccompanied." I said I wouldn't do that at night, and he stared at me meaningfully, noting that any time in downtown Nairobi on the street was very risky for a foreign visitor. A sobering lesson learned by this naïve American visitor. We were picked up by Mike Swift's driver and always accompanied in our trips around town.

As should be apparent by now, the milieu I was in at UGA fostered the simultaneous pursuit of several research and collaborative writing projects. During the few years of my tenure with Preston Hunter as Head of the Entomology Department, he gave me free rein to develop as many research projects as I could handle. This was too good to last, and when Arden Lea came in as the new head, he urged me to teach one or two seminar courses and to assist Dac Crossley in teaching the Insect Ecology course. Dac and I co-taught that course for at least six years until we moved into the Institute of Ecology full time as teaching and research faculty. Additional offerings included a course in stable and radioisotopes in ecology, a seminar in Soil Ecology and then beginning in 1989, a course in Soil Ecology. We built up some useful course notes, and within a few years, wrote up the first edition of our *Principles of Soil Ecology* book (Coleman and Crossley, 1996). By the mid-nineties, Dac and I invited our colleague Paul Hendrix to co-teach the course with us. After 1998, Paul and I handled a majority of the lectures, with Dac being a guest lecturer after his retirement. With the field moving along quite rapidly, we jointly authored a second edition (Coleman et al., 2004a). Small differences in packaging can affect book sales. In our case, the first edition was issued in hardcover, and only in paperback after 5 years had elapsed. For the second edition, we negotiated with Kelly Sonnack, the Biological Acquisitions Editor of Elsevier Academic Press for it to appear in paperback. The first year sales were more than triple of those of the first edition's initial year. It is gratifying to see the increased interest in the subject of Soil Ecology in current times. I invited Dr. Mac Callaham of the US Forest Service Athens Research laboratory to be a co-author with Crossley and me on a third edition of the textbook (Coleman et al., 2018).

38.
Studies in collaborative research: Some big projects that were unfunded

The Institute of Ecology was ready to tackle big things in the late 1980's under the direction of H. Ronald Pulliam, our director from 1987 onward for several years. A group of ecosystem researchers, including myself, thought we could apply successfully for a large interdisciplinary grant being offered under competitive terms to a few groups around the country during calendar 1989. Ron deputized Bud Freeman, a rising young star in the Georgia Museum of Natural History, and me to help lead such an effort. We christened it: "From the Mountains to the Sea: ecosystem studies along a continental gradient." We had several meetings of the entire UGA Ecology faculty, numbering more than twenty all told. Gene Odum, who had been suggesting this sort of project for at least the previous twenty years, but never had a critical mass of scientists to work on it, encouraged us avidly. In a sense, it took the sort of landscape-scale approach used by the Coweeta Long-Term Ecology Research project, and extended it across the Piedmont and down to the seacoast. We dealt with the complications of most of the runoff from Coweeta draining into the Little Tennessee and the Tennessee Rivers, and then into the Ohio and ultimately the Mississippi. However, that simply expanded our range, hydrologically, and was very intriguing to deal with in conceptual models.

After sweating away over the proposal and budget for an envisioned five-year project during spring and summer of 1989, we submitted it to the National Science Foundation and waited with bated breath for word from the program director, Dr. William Harris. Much to our chagrin, we found our effort made it through to the semi-finals, as it were, but we were beaten by a group of physicists and astronomers from the University of California,

Berkeley, who proposed to study dark energy and dark matter with their interdisciplinary team.

Ron Pulliam went on to try an even larger interdisciplinary study with scientists from Georgia Tech and UGA on aspects of biogeochemical cycling of radionuclides at the Savannah River site during 1990. The Department of Energy had millions of dollars to expend on site-based research. Somewhat unfortunately for our UGA group, the Tech researchers were more assiduous in developing close ties with the Savannah River Site funding sources, so the funds went virtually entirely to them. With the failed results of these extensive research group projects in mind, I resolved to wait for my "turn" to apply to be lead Principal Investigator of the Coweeta project when it came open in 1996; I discuss that experience later in this book.

39.
Studies in Agroecology: The Horseshoe Bend experiments

One of the greatest frustrations of soil ecology research was the difficulty in manipulating elements of the soil biota by either mechanical (screens) or chemical (biocides) means. We had a wonderful idea about setting up long-term experiments at several sites across the Georgia Piedmont, at Horseshoe Bend, Watkinsville and in Griffin, GA. We installed fine mesh screens in these sites to confine roots and measure root and soil-related nutrient uptake and organic matter dynamics, as well as using acaricides and vermicides to limit or curtail mites and worms in experimental 2 m2 plots.

This study, with Paul Hendrix as lead PI (Principal Investigator), was funded for five years. It was extensively replicated in the field, with a large team of grad students, including Petra Van Vliet and Shuijin Hu, and Mike Beare as postdoctoral, and several full-time and part-time technicians. We bought a van, and lots of analytical time for measuring the 13C and 15N-labelled materials that we grew in the experimental enclosures, following their fates over time.

We also had evening gatherings in the Fall months at Horseshoe Bend, accompanied with beer and sing-alongs.

(Left to Right) Dave Coleman, Barbara Mueller, Paul Hendrix and Dac Crossley. Ca. 1993.

LEFT PHOTO: Dac and Dave under a parasol at Horseshoe Bend on a July afternoon in the 1990's (photo courtesy Dac Crossley Jr.).
RIGHT PHOTO: (Left to right) Dave C., Preston Hunter and Dac Crossley. Occasion: A "Horseshoe Bender," keg party at the Horseshoe Bend site, Athens, GA. This celebrated our triple Scorpio birthdays (Hunter on Nov. 5, Crossley on Nov. 6 and Coleman on Nov. 7), ca. 1988.

One part of the study that was most fruitful was the detailed analyses of dynamics of water- stable macro- and micro-aggregates, which led to a series of papers with Mike Beare as lead author (Beare et al., 1994a, 1994b). These papers were widely cited in the international literature. However, the data generated by our group in the physical separation barrier study ended up with too large a variability to provide significant differences for the analyses of variance that we subjected our results to.

One of the more fruitful synthesis papers from this work was presented at the Soil Ecology Society meetings at Michigan State University in 1993 (Beare et al., 1995). Mike and our group developed the concept of "hot spots" in soil, including drilosphere, porosphere (Vannier, 1987), rhizosphere, detritusphere, and the one we emphasized, the aggregatusphere. Our thesis was that the relatively small volume of the soil influenced in these various ways constituted a series of "pressure points" in which the status or "health" of these systems could be studied most effectively. This was further elaborated in a major review with Paul Hendrix and Gene Odum as co-authors (Coleman et al., 1998).

Interest in agroecology has continued down to the present. In 2012 I served as a co-editor of a major synthesis volume, entitled: *Microbial Ecology in Sustainable Agroecosystems* (Cheeke et al., 2012).

Interest in plant-soil interactions has continued well into my retirement, now seventeen years' duration in summer of 2022. After presenting an invited paper to the Sixth Australasian Conference on Soil Pathogens in Twin Waters, Queensland, ca. 120 km. north of Brisbane in August of 2010, I wrote up a synthesis paper, entitled: "Understanding soil processes: one of the last frontiers in biological and ecological research" (Coleman, 2011). This was frequently cited at the time, and the publishers of the journal asked me to edit a synthesis volume on a similar topic. I asked a new generation to handle the task: John Dighton, Rutgers University, Camden campus, offered to edit such a book. He invited a colleague, Jennifer Krumins, of Montclair State University, to co-edit the book, entitled: *"Interactions in Soil: Promoting Plant Growth"* (Dighton and Krumins, 2014). The editors generously invited me to write a final chapter in the volume, Chapter 11 (Coleman et al., 2014), entitled: *"Toward a holistic approach to soils and plant growth,"* co-authored with my Chinese colleagues, Shenglei Fu and Weixin Zhang of Henan University and the South China Botanic Garden, in Guangzhou.

40.
Editorial Experiences I have had

Over the years I have served as a subject editor for several scientific journals. From 1978-81 I was editor for soils and invertebrates for the Ecological Society of America (*Ecology* journal), and later for *Soil Biology and Biochemistry* in 1980. John A. (Tony) Wallwork came through Fort Collins in 1982, and presented a seminar. He later nominated me to replace him as subject editor for *Pedobiologia*. I also edited manuscripts for *Applied Soil Ecology* and the *European Journal of Soil Biology*. The editorial duties that have been most extensive are on *Soil Biology and Biochemistry*. Some of our early work in Fort Collins was submitted to that journal when John Waid, the founding Editor-in-Chief, published it quarterly. The journal now appears monthly, publishes several hundred papers per year, and has the highest impact factor in its field. When John began receiving 8-10 manuscripts per day in 1997, he recruited Jo Anderson, Richard Burns, and me to join him as Chief Editors. By the early 2000s I edited over 200 articles per year, including initial receipt, and revision stages. John Waid passed away early in 2012, leaving a legacy of having founded the premier journal in soil biology and ecology. It continues to have the highest citation rating of any journal in its field. I inherited the mantle of Review Editor from John Waid in late 2012, continuing through early 2018.

41.
The influence of Scientists in East Asia on soil ecology

The big news in soil ecology as well as in all areas of science in the early 21st Century is the burgeoning activity in East Asia, including China, Taiwan, Japan and Korea. By far the greatest percentage increase has been in Chinese papers. The ideas are usually good, with the quality of the English language being quite variable yet, but improving markedly over the last several years. I made a scientific visit to Kunming, China, the Xishuangbanna Tropical Botanic Garden near the Laotian border, and to Beijing in April-May 2004. Basic research is given strong support by the Chinese Academy of Sciences (CAS) and the national government in general. Researchers in the CAS receive salary supplements of up to $4,000 US per paper accepted by an international journal.

Just prior to the trip to China, Barny Whitman from the UGA Dept. of Microbiology and I, with the assistance of Charles Chiu [Academia Sinica] and Hen-Biau King [Taiwan Forestry Research Institute [TFRI] and partial support from the Office of International Studies of the National Science Foundation held a five day symposium and workshop on "*Impacts of Soil Biodiversity on Biogeochemical Processes in Ecosystems.*" This meeting, on April 18-24, 2004, hosted by TFRI, drew over forty participants from North America, Asia, Australia and Israel. Many of the participants were members of the North American and Taiwanese LTER networks. We published a total of 17 manuscripts, several co-authored with Taiwanese scientists, which appeared as a special issue of Pedobiologia (Coleman and Whitman, 2005).

This collaborative effort grew out of an earlier scientific collaboration between Charles Chiu, Barny Whitman and me in the winter/spring of 2002 at UGA, when Charles was on a sabbatical half-year. I visited Charles and his colleagues at the Academia Sinica in August 2002, and was impressed by the nature and extent of their research work. Other funding from NSERC provided airfares for two Canadian colleagues to attend the 2004 meeting, and all local expenses were borne by Taiwanese agencies, principally TFRI (Director: Dr. Hen-Biau King) and the Research Center for Biodiversity of Academia Sinica (RCBAS). From 2004-2013 I served on the Academic Advisory Committee of RCBAS.

42.
Collaborative Research at its finest: The Long-Term Ecological Research Network (LTER)

One of the most intellectually challenging collaborations I have ever participated in is the Long-Term Ecological Research (LTER) program, funded by the NSF. I had watched with interest in 1980 as early plans were made for the establishment and operation of the CPER (formerly Pawnee site) LTER, which was renamed the Short-grass Steppe (SGS) LTER. Initial co-lead PIs were Bill Lauenroth and Indy Burke. With several other projects ongoing, I observed from the sidelines during the early 1980's and then became involved in the Coweeta LTER a few years after moving to Athens in 1985. Coweeta had just obtained a five-year renewal for the 1985-1990 period, so I began by taking a few students, namely Petra Van Vliet and Shuijin Hu, from the Horseshoe Bend project, up to Coweeta to sample soils on an elevational transect. Their work led to interesting contrasts between soils in lowlands (at Horseshoe Bend) and Coweeta, comparing Enchytraeids (small Oligochaetes related to earthworms, Van Vliet et al., 1995), and soil aggregation status and amounts of soil carbohydrates (Hu et al., 1997).

I invited John Dighton, then at the ITE Merlewood, U.K. research station, to spend three months with me at Coweeta in the summer of 1990. He sampled under rhododendron stands, measuring mycorrhiza distribution and abundances and phosphatase activities associated with the mycorrhizal roots. Surprisingly, he found three kinds of mycorrhiza on the rosebay rhododendron (*Rhododendron maximum*) roots, ectotrophic, endotrophic and ericaceous mycorrhiza (Dighton and Coleman, 1992). This diversity of mycorrhiza on one species was very unusual.

In the early 1990's, Coweeta was devoting considerable resources and personnel to setting up and instrumenting a series of five watersheds on an elevational transect from lowlands on Ultisols at ca. 750 m. elevation up to ca. 1450 m. on Inceptisols at a Northern Hardwoods site with 30% more precipitation. The instrumentation consisted of tension lysimeters, time-domain reflectometry (TDR) probes for soil moisture content, and numerous other measures of abiotic variables. The sites were set up with wooden walkways on the very steep hillslopes, and are still in use currently. Other graduate students, namely Karen Lamoncha working with Dac Crossley, and Randi Hansen with me, measured Oribatid mite abundances.

Karen sampled across all five transects, and found that the overall species richness was somewhat higher in the aggrading forested watersheds of Coweeta than in the old growth stands of the Joyce Kilmer natural area in the western end of North Carolina (Lamoncha and Crossley, 1998). Randi focused on the more speciose high-altitude (ca. 5,000 ft.) Northern Hardwood site, called Watershed 527, and conducted some innovative studies of physical and biochemical effects of leaf litter substrates on species richness in experimental enclosures (1 m2) within the 400-m2 study areas. Randi found significant decreases in species richness of Oribatids over time as she decreased the species richness of leaf litter from 5 down to 3 and then single species combinations of litter (Hansen and Coleman, 1998; Hansen, 1999, 2000). She linked this decrease to decreasing physical and chemical complexity of the litter substrates. She summarized her work in a by now classical synthesis paper in the "*Invertebrates as Webmasters in Ecosystems*" volume, entitled: "*Diversity in the Decomposing Landscape*" (Hansen, 2000).

One of the strong points in the LTER network in North America has been the extensive nationwide interactions facilitated by the Network Office. After a trial couple of years when there were only the six founding member sites, the network office moved to the University of Washington, under the capable supervision of Jerry

Franklin. The network grew markedly over ten years, and then the office moved again to the University of New Mexico, Albuquerque, who supported the extensive building-up of the Sevilleta LTER site, and the strong guidance of Jim Gosz. Beginning in 1997, the Network hired Bob Waide as Executive Director, which was a further improvement, with a critical mass of personnel to make contacts, etc. In the years under Jerry Franklin, the Network began a series of Synthesis Volumes related to sites and major ecological processes. The series was begun under Bruce Hayden, University of Virginia, Publications Committee chair. From 1997 onward, we encouraged the development of 10-12 more volumes. The first volume published was the *Tallgrass Prairie* synthesis edited by Alan Knapp and colleagues (1998).

Beginning in early 1996, a workshop was held at Sevilleta on the topic: "*Standard Soil Methods for Long-Term Ecological Studies*," encompassing all of the relevant physical, chemical and biological characteristics of interest. Phil Robertson from Michigan State University convened this interactive group of more than 35 scientists from all over North America. We produced drafts of 20 chapters, met again a few months later to critique them, and then sent them out for review to two or three peer reviewers.

We took the process very seriously, in some cases dropping non-performing authors, or adding authors to provide greater breadth to the chapters. This led to the publication of the volume (Robertson et al., 1999). It has been very well received and cited internationally. I enjoyed collaborating with three co-authors (John Blair, Ted Elliott and Diana Wall) to write the chapter on methods for studying soil invertebrates (Coleman et al., 1999).

The triennial All-Scientists' Workshop has been one of the most scientifically rewarding aspects of the LTER Network activities. This began in 1984 at Lake Itasca, Minnesota, and then several meetings were held at the meeting center of the YMCA of the Rockies at Estes Park, Colorado. It was held in 2000 in conjunction with the ESA meetings at Snowmass, Utah, and in 2003 at Seattle, Washington. The hallmark of the meetings is a large number of poster sessions that draw people out and encourage considerable interaction and discussion. A number of synthesis workshops have been held during and after the meetings, and has led to some continuing collaboration on cross-site synthesis studies. As an example of the level of enthusiasm generated, Ariel Lugo of the USFS, Puerto Rico attended the entire session on decomposition dynamics at the 2003 meeting because the combination of field data and mathematical models was so attractive to him. The trio of young scientists who organized it, Drs. Grizelle Gonzalez, Jean Lodge and Whendee Silver, typify the capabilities of the rising young generation who are carrying on further research in the LTER Network.

43.
LTER Site Reviews, a great learning experience

I participated in several site reviews of LTER sites. One of the more noteworthy ones was in the summer of 1991. July 1, I met an NSF review team, led by Tom Callahan, at Fairbanks, Alaska to review two LTER sites: the Taiga LTER and the Tundra project at Toolik Lake, north of the Arctic Circle. It was an all day drive up the Haul Road. There were many interesting terrestrial and aquatic ecology projects. It was my first experience of the 24 hr. daylight at Toolik. Such a vast landscape; we felt immersed in very big ecosystems, the ultimate Big Sky country. The investigators worked very long days, from 18 to 20 hours per day, attempting to complete several experiments in just six weeks or more time.

Flights to and from Alaska took three steps on Boeing 727s: Atlanta to Cincinnati and then Seattle, and on to Fairbanks. Travel time was 12 hours each way.

44.
Experience with NSF Proposal Reviews

In the 1980's and early 1990's I served on Panels in the NSF reviewing, first, Ecology proposals, and then later on, Ecosystem proposals. We received from twenty to twenty-five proposals, and took the lead on a half dozen or so and served as co-reviewers in the remainder. This was a wonderful learning experience, comparing notes with colleagues from all over the country on their impressions of where our fields of inquiry were going. In the 1980's we wrote out our panel summaries longhand to be typed in by secretaries in the NSF. By the 1990's we typed in our review summaries directly into electronic files. An interesting time of transition, for sure.

45.
Graduate student advising experience

I have advised over 35 graduate students and 7 postdoctorals in my career, and served on the advisory committees of at least two hundred more. Many of my students went on to publish numerous research papers in internationally-refereed journals. One of the early ones to pursue a different path was James T. (Tom) Callahan. Tom was clearly cut from a different cloth, was widely read, and possessed of a virtually photographic memory to boot. I advised Tom for his Master's and, after I moved to Colorado State University in early 1972, he finished the last six months of his Ph.D. degree with Dac Crossley to complete his dissertation on the population ecology of fall webworms at the Savannah River site. Tom interviewed with the NSF Environmental Sciences Division in early 1972, and was hired on to be the permanent member of the Ecosystem Sciences directorate. Senior scientists like Paul Risser, Jerry Franklin and Wayne Swank, were "rotators," with a typical service time of two years. Tom stayed on, and saw the program through some lean times (several years of flat funding—no increases in funds) in the late 1970's and most of the 1980's as well. His lifetime publication record was rather short, but two of his papers (Callahan 1984, 1991) document his lifelong support for the LTER Network. This is a strong legacy for posterity indeed.

A number of us senior scientists in LTER organized a "surprise" luncheon testimonial gathering to thank and "roast" Tom at the Ecological Society meetings in Albuquerque in the summer of 1997. He was delighted at the recognition. Little did we realize that he had a combination of lung and bone cancer beginning at that time, and after much chemo- and radiation therapy, he died two years later in 1999.

A major motivating factor bringing together ecologists and soil scientists in the 1990's was the concept of "soil quality." Beginning with scientific meetings of the Soil Science Society of America in 1992, and continuing through 1996, John Doran et al. held a series of symposia on this general topic. I served as coeditor with John on a volume, entitled: *"Defining Soil Quality for a Sustainable Environment"* (Doran et al, 1994). I served as co-author with Dennis Linden, USDA St. Paul MN research laboratory on the role of soil fauna in soil quality (Linden et al, 1994). This was followed up two years later with a volume entitled: "Methods for Assessing Soil Quality (Doran et al., 1996)." As with all innovative and easy to misinterpret concepts, this series of meetings and papers met with considerable resistance. The purists in soil science pointed out quite rightly that quality is difficult to measure, and is also subject to misinterpretation. Interestingly, a major impetus for these meetings and books was strong pressure from state and federal agencies that were charged with developing metrics for measuring land use and ways in which lands are degraded through overgrazing, unsuitable cultivation, etc. The requirements for Best Management Practices (BMPs) have been increasing over the last decades. As noted earlier above, this work with John Doran and other soil scientists led to a major review paper in another Soil Science Society of America special volume (Coleman et al., 1998).

The majority of the research reviewed in the papers cited in the previous paragraph was in agricultural ecosystems. Our research group in the Coweeta LTER proceeded to ask the question: what sorts of biological indices of soil quality are both informative and useful in forest ecosystem studies? This led to a major synthesis effort coordinated by Jennifer Knoepp, the forest soils and nitrogen cycling authority in our project. After a few meetings of co-authors, Jennifer presented her results in national scientific meetings, resulting in a synthesis paper entitled: *"Biological indices of soil quality: an ecosystem case study of their use"* (Knoepp et al., 2000). The gist of the paper is that a suite of biological and chemical characteristics is required to adequately characterize the functional ecology of a given site. These values change over time and changes in climate. They depend on the use or goal of the ecosystem under study.

Another study long in gestation had arisen from our interests in possible linkages between uplands and aquatic systems. Similar concerns had led Christien Ettema, a second Dutch graduate student working with me, to study riparian processes in pasture and forest lands in south Georgia on a dairy farm near Tifton, GA. Receiving EPA funding support, she made some of the first measurements documented of changes in nematode populations in a series of ecosystems that were nitrogen-stressed (Ettema et al., 1998, 1999).

To conduct research on hillslope-riparian processes, our Coweeta group with Wayne Swank, Judy Meyer, Alan Yeakley and I in the lead, established two study sites in an area of high relief near the northeastern boundary of the Coweeta Hydrologic Laboratory lands. Called Watershed 55, it has slopes of 60-70%, lush rhododendron understory, and tulip poplar and southern red oak overstory. We chose instrumented study areas of 10 x 15 m. with lysimeters, Time-domain reflectometry (TDR) rods, and wells in the stream to measure nutrient movement within the small watershed. The control area was upstream from the lower site, both of the study sites having northeastern exposures. The experimental treatment was a complete extirpation of the rhododendron understory by hand and chain saws in late August 1995, with all of the material hauled offsite by hand.

The control site quickly became a second treatment site, with the onslaught of Hurricane Opal in early October of that year. After more than 25 cm. of intense rainfall in less than two days, Opal hit several parts of Coweeta (including our control site) with microbursts of high winds, knocking down seven large old growth trees, mostly southern red oak (*Quercus falcata*), one of which was 270 years old. Our initial hypothesis of significant downhill movement in an undisturbed hillslope land with no rhododendron was not supported, but the effects of large tip-up mounds from the windthrows in the uphill study site on inorganic N, sulfate, carbonate and other anions was very large and gradually diminished through the soil profile (Yeakley et al, 2003). In sum, the pedoturbation of the natural disturbance far outweighed (by a factor of 103) the impacts of the understory clearcutting.

In retrospect, we needed a disturbed surface treatment with the understory removal, but limited personnel and support budgets precluded any expansion of the study. By studying the same sites several years before, during and after treatment, we dealt forthrightly with the problem of pseudoreplication, using randomized intervention analysis (RIA) developed a few years before by Steve Carpenter of the University of Wisconsin. The before-treatment studies of soil ecology were presented by Maxwell and Coleman (1995); the during- and after-treatment results by Wright and Coleman (2002).

46.
Ecological research projects in the American tropics

Conducting significant amounts of research in tropical regions remained a major lacuna in my scientific life until the 1990's. With the arrival of Ron Carroll as Associate Director of the Institute of Ecology in the late 1980's, a greater emphasis on tropical studies began for several students and faculty, including me. Dac Crossley and I visited Xiaoming Zou at the Luquillo site in Puerto Rico in October of 1992, and saw that there was considerable promise for some collaborative work. Dac and Jean Lodge of the USFS in Puerto Rico had proposed some pioneering studies of litter decomposition and fungal diversity in a wide range of leaf litters at the El Verde site in the early 90s. These required considerable manpower and supplies that were not approved by the NSF review panels, leading to "declinations" (as they are called by NSF) of such proposals. I suggested to Dac that we try a streamlined approach, focusing on hiring a postdoctoral fellow to conduct a comparative study of microarthropods and their involvement in leaf litter decomposition in temperate and tropical ecosystems. We submitted this to the Cross-site Studies section of *Ecosystems* in NSF in late 1993, and it was funded in midyear 1994.

Liam Heneghan, a newly-minted Ph.D. from Tom Bolger's group at University College Dublin, be-

(Left to right) Ron Carroll, D. Coleman, and Amy Rosemond. 1990's (Photo courtesy UGA Odum School of Ecology)

gan his postdoctoral work with us in September 1994, and by early 1995 our group, including Liam, Xiaoming, Dac, Bruce Haines and I had a comparative field study in place, following litter decomposition of Chestnut oak Quercus prinus at Coweeta, the Luquillo site and also on the Atlantic coastal area of La Selva in Costa Rica.

The logistics of setting up study plots that included addition of naphthalene crystals on litter bags to serve as non-arthropod controls were daunting, but our colleagues in Puerto Rico and Costa Rica were most helpful. The end results were that, although microarthropods caused significant increases in litter decomposition in all sites, the relative impacts of the more speciose Oribatid species were greatest in the La Selva site, with Luquillo being intermediate and Coweeta microarthropods having the least marked effect (Heneghan et al., 1998, 1999a, 1999b). In other words, the tropical site had the greatest impact on rates of decomposition.

Nearly contemporaneous with the preceding study, Ron Carroll and I began exploring the possibility of studying plant-soil interactions with Mike Miller of the Ecology group at Argonne National Laboratory. We were particularly interested in the roles of mycorrhiza and soil fauna in revegetation processes in the montane Andes region of Ecuador. Mike knew a colleague in the MacArthur Foundation office in Chicago, who encouraged us to submit a proposal based on the topics noted above.

We were funded for the years 1994-96, for travel to Quito and research on site by two Ph.D. students, Greg Eckert, working on mycorrhiza with Ron Carroll, and Chuck Rhoades, studying nitrogen and carbon dynamics with me. These studies were in successional seres of forest and grassland areas (elevations 1200-1600 m.) on Andosolic soils of the Thomas Davis Station of the Fundación Maquipucuna.

Rodrigo Ontañeda and his wife, Rebeca Justicia, president and general managers of the Foundation, facilitated our research. Greg and Chuck's studies, subsequently published as Rhoades et al. (1998, 2000), among others, were the forerunners of a series of research projects at the Davis station that continue to the present. Guanglong Tian, a research associate with our Soil Ecology laboratory from 2002-2004, carried out some follow-up studies at the Ecuadorian site to compare the deposition of carbon under bamboo (*Guadua* sp.) stands and in nearby grassland and forested area. The bamboo stands sequester significant amounts of carbon, similar to the *Paspalum* grasslands nearby (Tian et al., 2007).

The pasture grasses had been introduced by well-meaning US Agency for International Development (US/AID) officials in the 1960's to encourage raising dairy cattle as an alternative food source, but the grasses proved to be of low nutrient quality and of little use for dairy farming. The intent had been to wean the indigenous people from growing sugar cane and producing local "white lightning" liquor to sell in the markets of Quito. Bamboo is a widely used item in home construction, and more widely used than regular lumber in building homes in the local communities of montane Ecuador.

Ron Carroll and colleagues have proceeded to experiment with the growth of shade-grown coffee in the areas around Maquipucuna. They have persuaded a local coffee house in Athens, GA to market the Ecuadorian shade-grown coffee in the Student Learning Center and at other locations on the UGA campus, providing an income to the farmer/growers three times greater than they could get by selling to coffee wholesalers in Quito.

In addition to collaboration on various synthesis and research projects in the New World, the opportunity for international collaboration has always been strong in the ecosystem community. Beginning with some workshops on agroecosystem dynamics led by Lijbert Brussaard and his group at the Instituut voor Bodemvruchtbaarkeid (IB) or Institute of Soil Fertility in the north of Holland in 1988 (including a large contingent from the NREL, such as Vern Cole, Bill Parton and Ted Elliott), these synthesis meetings continued into the 1990's, with one being convened by Lijbert in Wageningen on Biological Factors in Soil Structure in November 1991.

Years later, Diana Wall, as part of the Soils and Sediments diversity section of the International Biodiversity Observation Year (IBOY) convened a series of workshops in Lunteren, the Netherlands, in October 1998, and Corvallis, Oregon in August 1999. A wide range of people involved in the synthesis papers produced several papers: Hooper et al. (2000), and Wolters et al. (2000). The cast of characters changed somewhat with each meeting, with one of the more notable synthesis papers from Corvallis being written by Bardgett et al. (2001). An equally star-studded cast was provided by a wide range of freshwater and marine aquatic ecologists, as can be seen by the papers in the appropriate issues of BioScience and Ecosystems. All this work culminated in a large SCOPE volume, edited by Diana Wall (2004), entitled: "*Biodiversity in Soils and Sediments.*"

Diana convened a larger group of editors and authors in 2011 and produced a landmark volume, entitled "*Soil Ecology and Ecosystem Services*" (Wall, et al., 2012).

47.
Branching out in the early phases of my retirement, 2006 onward

I officially retired in June of 2005, and several colleagues, notably Janice Sand in the Odum School of Ecology, arranged a really major celebration. With help from Dr. Gordhan Patel, the Vice President for Research, Janice was able to provide air fare for more than twenty colleagues from Europe and Australia, plus several of my former grad students to attend a gala reunion celebration. We had a day-long symposium with lots of discussion. It was published as a special issue of Pedobiologia (volume 50 (6), 4 Jan. 2007).

At the "Davefest," Dave's retirement celebration, October 2005, at Flinchum's Phoenix, Athens, GA. (Left to right) Breana Simmons, Dave and Dac's secretary Lindalee Enos, Herman Verhoef (Vrije Universiteit, Amsterdam, the Netherlands), Fran Coleman, Dac Crossley, Mitch Pavao-Zuckerman, and Stephanie Madson. Breana, Mitch and Stephanie were Dave's Ph.D. students.

A few months after my retirement, I was invited to spend three months (February through April 2006) on a Senior Research Fellowship with colleagues in the CSIRO Division of Entomology, later merged into the Division of Ecosystem Studies. My host for this visit was Gupta Vadakattu, a longtime friend and colleague who has spent a lifetime of research on the microbial and faunal ecology of dryland (i.e., rain-fed, unirrigated) agricultural systems. I followed Gupta's research career with interest, dating from his Ph.D. Dissertation in 1988 on soil food webs in agroecosystems in the wheat lands of Saskatchewan. I served as his external reviewer at the time he defended his dissertation in the early spring of 1988, with Jim Germida as his major professor and John W.B. Stewart as Dean of the College of Agriculture at the University of Saskatchewan, Saskatoon.

(Left to right) Gupta Vadakattu, Dave Coleman and a New South Wales botanist at cotton field site, Narrabri, NSW, Australia, March 2006.

The CSIRO grants two or three Senior Fellowships per year, named the McMaster Visiting Fellowships. With Gupta's encouragement, I applied in December and the funding was granted just a few weeks later. The stipend covered all travel costs from USA to South Australia and return, lodging and incidental costs, and extensive internal travel within the country. In addition to advising on research work in Adelaide on the Waite Campus in the suburb of Glen Osmond, I presented seminars in Canberra, at CSIRO headquarters, and also in Narrabri, a small town with a research field station associated with the Cotton Research Centre, as well as in Adelaide.

I took my wife Frances and we got to experience a South Australian late summer and early autumn in the time frame of February through April 2006. We stayed first for a few weeks in an apartment in an old part of town, Kensington, dating from the early settlement days in the 1840's, and later for a final seven weeks in Henley Beach, along the Gulf of St. Vincent, a lovely estuarine area to the west of downtown Adelaide. I was impressed by the frequency with which scientists traveled over long distances across country to attend research workshops and scientific meetings. Bearing in mind the total number of Ph.D. level scientists, some forty thousand nationwide, the travel budget was quite large. QANTAS, the national airline, has an impressive fleet of Boeing 737s and smaller turboprop planes to cover a wide array of small and medium-sized airports around that big country. Gupta and I developed a research study centered on some of his long-term research plots in Narrabri in central New South Wales, and not far outside Adelaide as well. We asked the question: how do detrital food webs function in the heavy clay Vertisols of the interior of Australia, and how different are they from similar food webs in the continental USA?

In the late 1980's, Gupta had applied for a postdoctoral research fellowship with CSIRO in Canberra soon after obtaining his Ph.D. degree, and I had written a strong letter of recommendation for him to his higher-ups. Apparently they informed him that I gave him about as strong a recommendation as they had ever seen, so they would have similarly high expectations of his future performance. Judging by his very impressive research output and productive career overall, our expectations were borne out very well. Being about as far away from home as it is possible to get, I rediscovered the truth of the old adage: "an expert is someone who is a long way from home." There were uniformly large and attentive audiences at the several seminars I presented around the country. It was also good to visit old friends, now long retired, such as Drs. Ken Lee and Albert Rovira in Adelaide and also with Prof. John Waid and his wife Sally with whom we spent a charming day at their home near Mooloolaba, about 70 km. north of Brisbane, in Queensland.

One of the side benefits of such a senior sabbatical, as it were, was the chance to see numerous sights in the wonderfully varied landscape that constitutes the very large island continent of Australia. For sightseeing, we traveled to the South Australian coast and visited the Coorong delta lands, where the Murray River empties into the South Pacific Ocean, with its thousands of water birds, including the amazingly large Australian pelicans. On our way to the Coorong, we saw literally hundreds of sulfur crested Cockatoos hanging out in trees near the small towns we traveled through. We also took a long weekend to visit the famous Kangaroo Island, a wildlife sanctuary about 30 miles off the South Australian coast.

We went there in mid-March, and

Large Australian pelicans in the Coorong Estuary, March 2006.

saw the rapid change of seasons, with the first two days warm and summer-like. Then, almost as if a large switch had been flipped, cold and dry air from the Antarctic began flowing in, and we suddenly needed sweaters and jackets to make the trip back north to Adelaide. Toward the end of our visit, in mid-April, we drove for several hours due north of Adelaide to the ancient Flinders Range, where we saw more dryland vegetation and had a chance to travel on the Pichi Richi Railway, a remnant narrow gauge line that once connected to a railroad to Alice Springs and farther north.

48.
Major synthesis publications in the second decade of the New Millennium

I was invited to participate in a synthesis volume celebrating the fiftieth anniversary, in 2017, of the founding of the Natural Resource Ecology Laboratory at CSU. We first met in September of 2015, and worked up outlines of our chapters. This book had thirteen chapters and dozens of authors. Bob Woodmansee cajoled and encouraged (led) us masterfully. Our chapter was on the evolution of ecosystem science (Coleman et al., 2021, in Woodmansee et al., 2021).

Credit: Cambridge University Press

49.
My viewpoints on Global Climate Change

*M*uch of my research work preceded major concerns about global climate change. It is apparent from work over the last 10-20 years that climate change and humans' influence on it is accelerating rapidly. I hope that climate change deniers, such as a recent past-president of the US, must not be given the reins in a future election over such a consequential set of biogeochemical changes now in existence.

Philip Corbett, a noted British ecologist, as early as 1970, called man the "earth pest," prophesying what the next fifty years would look like with virtually unrestrained population growth and massive increases in fossil fuel usage over that period. Decades ago, Howard T. Odum (1971) addressed the false economies of switching over to nuclear power to save carbon inputs. He made a cost accounting of the additional costs of maintaining and containing nuclear wastes, noting that full accounting for these costs over thousands of years was costly indeed.

50.
A retrospective look at major persons and personalities who influenced me

Looking back on my forty plus years of interactions with colleagues in many countries across a range of disciplines, two people stand out. Firstly, Amyan Macfadyen, with his indomitable will power and very perceptive nature about how ecosystems function. A co-dominant with him is Gene Odum, who influenced the Institute of Ecology even in his 17-plus years of retirement by sheer force of ideas and will power. Gene was a mentor to me in my early years at the Savannah River Ecology Laboratory.

The anecdotes about him are legion, and a few of them have made it into the pages of the masterful biography of him written by Dr. Betty Jean Craige (2001) on the UGA campus. Although Gene was an avid proponent of having a strong core of basic research, he conducted a one-man outreach effort before some of us were even aware of impending environmental problems.

One example of this was soon after Lester Maddox was elected governor of Georgia in 1966. There was a strong movement afoot by outside investors to buy up marshlands along the Georgia coast for phosphate mining, somewhat akin to the extensive inroads that had been made along the coast of Louisiana. Gene went over to Atlanta and testified to the Georgia legislature about the need to maintain the marshlands and barrier islands intact, so the estuaries would continue to serve as nurseries for the extensive shrimp and other fisheries along the coast.

Governor Maddox invited Gene to accompany him and a group of legislators on a chartered airplane to view the coastal lands firsthand, because there were literally hundreds of millions of dollars involved in the "development" process. Gene, Lester and others donned hip boots and waded out into the salt marshes at low tide and viewed oyster beds, and the life teeming in them. Maddox was skeptical about how much the mining would impact what already seemed to be an inherently dirty, muddy system.

Odum pointed out that the phosphate mining by the outsiders, who he called "carpetbaggers" (a highly inflammatory phrase then as now), would turn the oyster beds into "oyster ghettos." Maddox, ever one to use a catchy phrase, said: "See, boys, the mining will make oyster ghettos. We can't have that, can we?" The move to sell off the lands for phosphate mining was stopped soon afterward. We in Georgia have many less-disturbed coastal areas as a result of Gene Odum's strong advocacy and colorful phraseology.

In conclusion, it has been an eventful and rewarding life in the ecosystem fast lane. I was able to "develop my own niche," as E.P. Odum commented to me when I assumed a full professorship position in the Institute of Ecology here at UGA, and pursue a variety of interdisciplinary studies of ecological interactions in the belowground part of terrestrial ecosystems. Going from the era of "through a ped darkly" (Coleman, 1985) to the analyses of species diversity in soil systems ("from peds to paradoxes") (Coleman, 2008), my scientific and life's journey has been one of numerous surprises and rewards along the way.

While at UGA from the mid-1980's onward, I occasionally traveled to Colorado to visit and work with friends (Diana Wall, Dave Swift, et al) at the NREL at CSU. I often visited with my sister, Margery Ferguson, and her two daughters, Katie and Tina, in Denver. On a trip in 1995, I visited with Margery at her favorite restaurant in Lyons, northwest of Denver. Sometimes we visited with Carol and Ric Tarr in Lakewood, west of Denver's downtown.

THE WORLD TRAVELS OF DAVE COLEMAN

Courtesy Google Maps © 2022

COUNTRIES DAVE COLEMAN TRAVELED

Italy—Naples, Rome, Florence, Milan, Padua, Venice
France—Paris, Rouen, Lyon, Rheims, Montpellier
Great Britain—London, Swansea, Bangor, Cardiff, Wales. Grange-over-Sands,
    York, Belfast, Coleraine, Dorset, Exeter, Oxford, Cambridge.
    Edinburgh and Aberdeen, Scotland
Ireland—Dublin, County Clare, Rosslare
Sweden—Stockholm, Uppsala
Finland—Helsinki, Jyvaskyla
Russia—St. Petersburg, Moscow
Denmark—Copenhagen, Aarhus
Poland—Warsaw, Poznan, Krakow
East Germany—East Berlin
Germany—Frankfurt, Giessen, Mainz, Wiesbaden, The Black Forest,
    Duesseldorf
The Netherlands—Lunteren, Groningen, Wageningen, Amsterdam
Belgium—Brussels, Louvain
Spain—Madrid, Zaragosa, Cordova
Switzerland—Zurich, Bern, Lausanne, Chur, Brig
Czech Republic—Prague, Ceske Budejovice, Brno
Austria—Vienna
Yugoslavia—Split, Zagreb in Croatia, Belgrade, Serbia
Greece—Athens, Isle of Hydra, Delphi
Israel—Jerusalem, Eilat, Tel Aviv
India—New Delhi, Varanasi, Agra, Lucknow, Mumbai
Thailand—Bangkok, River Kwai
China—Beijing, Kunming, Shanghai
Taiwan—Taipei, Kaohsiung, Chinmen
Japan—Osaka, Kyoto
Africa—Nairobi and Kisumu, Kenya; Kampala, Uganda, and Dakar, Senegal,
    Monrovia, Liberia and Lagos, Nigeria in transit only
Australia—Sydney, Adelaide, Narrabri, New South Wales, Kangaroo Island off
    the S. Australia coast, Melbourne, Brisbane, Darwin, Alice Springs
New Zealand—Auckland, Wellington, Hamilton, Palmerston North,
    Queenstown and Dunedin on South Island
South America—Brazil, Ecuador
Central America—Costa Rica, Panama, Belize
North America:
    Mexico—Mexico City, Acapulco, Ensenada, Tijuana, Tecate
    USA—all fifty states, incl. Alaska and Hawaii
    Canada—British Columbia, Alberta, Saskatchewan, Manitoba, Ontario,
        Quebec, Nova Scotia, New Brunswick

REFERENCES

Bardgett R.D., J.M. Anderson, V. Behan-Pelletier, L. Brussaard, D.C. Coleman, C. Ettema, A. Moldenke, J.P. Schimel, and D.H. Wall. 2001. The influence of soil biodiversity on hydrological pathways and the transfer of materials between terrestrial and aquatic systems. *Ecosystems* 4: 421-429.

Beare, M.H., P.F. Hendrix, and D.C. Coleman. 1994a. Water-stable aggregates and organic matter fractions in conventional and no-tillage soils. *Soil Science Society of America Journal* 58: 777-786.

Beare, M.H., M.L. Cabrera, P.F. Hendrix, and D.C. Coleman. 1994b. Aggregate-protected and unprotected pools of organic matter in conventional and no-tillage soils. *Soil Science Society of America Journal* 58: 787-795.

Beare, M.H., D.C. Coleman, D.A. Crossley, Jr., P.F. Hendrix and E.P. Odum. 1995. A Hierarchical approach to evaluating the significance of soil biodiversity to biogeochemical cycling. *Plant & Soil* 170: 5-22.

Blair, W. F. 1977. *Big Biology. The US/IBP.* Dowden, Hutchinson & Ross, Stroudsburg, PA.

Breymeyer, A. I., and G. M. Van Dyne (Eds.) 1980. *Grasslands, systems analysis, and man.* Cambridge University Press, Cambridge.

Callahan, J. T. 1984. Long-Term Ecological Research. *BioScience* 34: 363-367.

Callahan, J. T. 1991. Long-term Ecological Research in the United States: a Federal perspective. Pp. 9-21 in: Risser P. G. (ed.) *Long-term Ecological Research. An international perspective.* SCOPE 47, Chichester, U.K. John Wiley & Sons.

Chapin, F.S. III. 1980. The mineral nutrition of wild plants. *Annual Review of Ecology and Systematics* 11: 233-260.

Chapin, F.S. III, E.D. Schulze and H.A. Mooney. 1990. The ecology and economics of storage in plants. *Annual Review of*

Ecology and Systematics 21: 423-447.

Cheeke, T., D.C. Coleman, and D.H. Wall (Eds.) 2012. *Microbial Ecology in Sustainable Agroecosystems.* CRC Press, Boca Raton, FL.

Cheng, W., and D.C. Coleman. 1990. Effect of living roots on soil organic matter decomposition. *Soil Biology & Biochemistry* 22: 781-787.

Cheng, W., D.C. Coleman, C. R. Carroll, C. A. Hoffman. 1994. Investigating short-term carbon flows in the rhizosphere of different plant species using isotopic trapping. **Agronomy Journal 86: 782-788.**

Cheng, W., Q. Zhang, D. C. Coleman, C. R. Carroll and C. A. Hoffman. 1996. Is available carbon limiting microbial respiration in the rhizosphere? *Soil Biology and Biochemistry* 28: 1283-1288.

Clark, F.E., and E.A. Paul. 1970. The Microflora of grassland. *Advances in Agronomy* 23: 375-435.

Clark, F. E., and D. C. Coleman. 1972. Secondary productivity belowground in Pawnee Grassland, 1971. *US/IBP Grassland Biome Tech. Rep. No. 169.* Colorado State Univ., Fort Collins. 23 p.

Cole, C. V., G. S. Innis, and J. W. B. Stewart. 1977. Simulation of phosphorus cycling in semiarid grasslands. *Ecology* 58: 1-15.

Coleman, D. C. 1966. The laboratory population ecology of *Kerona pediculus* (O.F.M.) epizoic on Hydra spp. *Ecology* 47: 705-711.

Coleman, D. C. 1968. A method for intensity labelling of fungi for ecological studies. *Mycologia* 60: 960-961.

Coleman, D. C. 1971. Nematodes in the litter and soil of El Verde rain forest, Chapter E-7, p. E-103-E-104. In H. T. Odum, and R. F. Pigeon (Eds.) *A Tropical Rain Forest: A study of irradiation and ecology at El Verde, Puerto Rico.* USAEC TID-24270.

Coleman, D. C. 1973 a. Soil carbon balance in a successional grassland. *Oikos* 24: 195-199.

Coleman, D. C. 1973 b. Compartmental analysis of "total soil respiration": An exploratory study. *Oikos* 24:361-366.

Coleman, D. C. 1985. Through a ped darkly—an ecological assessment of root soil-microbial-faunal interactions. pp. 1-21 In A. H. Fitter, D. Atkinson, D. J. Read, and M. B. Usher (Eds.), *Ecological Interactions in the Soil: Plants, Microbes and Animals.* British Ecological Society Special Publ. 4. Blackwells, Oxford.

Coleman, D.C. 2008. From Peds to Paradoxes: linkages between soil biota and their influences on ecological processes. *Soil Biology & Biochemistry* 40: 271-289.

Coleman, D.C. 2010. *Big Ecology: the Emergence of Ecosystem Science.* University of California Press, Berkeley, CA.

Coleman, D.C. 2011. Understanding soil processes: one of the last frontiers in biological and ecological research. *Australasian Plant Pathology* 40: 207-214.

Coleman, D.C. and D. A. Crossley, Jr. 1996. *Fundamentals of Soil Ecology.* Academic Press, San Diego, CA.

Coleman, D.C. and B. Fry (Eds.). 1991. *Carbon Isotope Techniques in Plant, Soil, and Aquatic Biology.* Academic Press, San Diego.

Coleman, D. C., and A. Macfadyen. 1966. The recolonization of gamma-irradiated soil by small arthropods. *Oikos* 17: 62-70.

Coleman, D. C., and J. T. McGinnis. 1970. Quantification of fungus—small arthropod food chains in the soil. *Oikos* 21: 134-137.

Coleman, D. C., and A. Sasson (coords.). 1980. Decomposers subsystem, Chapter 7, p. 609-655. In A. Breymeyer and G. Van Dyne (Eds.) *Grasslands, systems analysis, and man.* IBP Synthesis Vol. 19. Cambridge Univ. Press, London.

Coleman, D.C., and W.B. Whitman. 2005. Linking species richness, biodiversity and ecosystem function in soil systems. *Pedobiologia* 49: 479-497.

Coleman, D. C., C. P. P. Reid, and C. V. Cole. 1983. Biological strategies of nutrient cycling in soil systems. *Advances in Ecological Research* 13:1-55.

Coleman, D.C., J.M. Oades and G. Uehara (Eds.). 1989. *Dynamics of Soil Organic Matter in Tropical Ecosystems.* University of Hawaii Press, Honolulu. 249 p.

Coleman, D.C., P.F. Hendrix, and E.P. Odum. 1998. Ecosystem Health: An Overview. pp. 1-20 in: P.H. Wang (Ed.) Soil Chemistry and Ecosystem Health. *Soil Science Society of America Special Publication* No. 52, Madison, Wisconsin.

Coleman, D.C., M. A. Callaham, and D.A. Crossley, Jr. 2018. *Fundamentals of Soil Ecology*, 3rd Edn. Elsevier/Academic Press. 369 pp.

Coleman, D.C., D.A. Crossley, Jr., and P.F. Hendrix. 2004a. *Fundamentals of Soil Ecology*, 2nd Edn. Elsevier Academic Press, San Diego, CA. 384p.

Coleman, D.C., D. M. Swift, and J. E. Mitchell. 2004b. From the Frontier to the Biosphere: a brief history of the USIBP Grasslands Biome program and its impacts on scientific research in North America. *Rangelands* 26: 8-15.

Coleman, D.C., W. Zhang and S. Fu. 2014. Toward a Holistic Approach to Soils and Plant Growth. Chapter 11, pp. 211-223 in: *Interactions in Soil: Promoting Plant Growth*, edited by Dighton, J., and Krumins, J. Springer Dordrecht, Heidelberg and New York.

Coleman, D. C., J. M. Blair, E. T. Elliott, and D. H. Wall. 1999. Soil Invertebrates, pp. 349-377 in: G. P. Robertson, D. C. Coleman, C. S. Bledsoe, and P. Sollins, Eds. *Standard Soil Methods for Long-Term Ecological Research*, LTER Network Series. Oxford University Press, New York.

Coleman, D.C., A. L. Edwards, A. J. Belsky and S. Mwonga. 1991. The distribution and abundance of soil nematodes in East African savannas. *Biology & Fertility of Soils* 12: 67-72.

Coleman, D. C., J. E. Lloyd, R. J. Lavigne, and A. Breymeyer.

1973. Feeding activities of soil macroarthropods at the Pawnee Site, 1971. *US/ IBP Grassland Biome Technical Report No. 215.* Colorado State Univ., Fort Collins. 9 p.

Coleman, D.C., E. A. Paul, S. Lynn and T. Rosswall. 2021. Evolution of ecosystem science to advance science and society in the 21stcentury. Chapter 3 in: R.G. Woodmansee, et al. 2021.

Craige, B.J. 2001. *Eugene Odum: Ecosystem Ecologist and Environmentalist.* University of Georgia Press, Athens, GA.

Crapo, N. L., and D. C. Coleman. 1972. Root distribution and respiration in a Carolina old field. *Oikos* 23: 137-139.

Dighton, J.D. and D.C. Coleman. 1992. Phosphorus relations of roots and mycorrhizas of *Rhododendron maximum L.* in the southern Appalachians, N. Carolina. *Mycorrhiza* 1: 175-184.

Dighton, J., and J. Krumins (Eds.). *Interactions in Soil: Promoting Plant Growth.* Springer Dordrecht, Heidelberg and New York.

Doran, J.W., D.C. Coleman, D.F. Bezdicek and B. A. Stewart (Eds.) 1994. *Defining Soil Quality for a Sustainable Environment.* SSSA Special Publication Number 35, Madison, Wisconsin.

Doran, J.W., and A.J. Jones (Eds.) 1996. *Methods for Assessing Soil Quality.* SSSA Special Publication No. 49, Madison, Wisconsin.

Dyer, M.I., M.A. Acra, G.M. Wang, D.C. Coleman, D.W. Freckman, S.J. McNaughton and B.R. Strain. 1991. Source-sink carbon relations in two *Panicum coloratum* ecotypes in response to herbivory. Ecology 72: 1472-1483.

Elliott, E. T., and D. C. Coleman. 1977. Soil protozoan dynamics in a shortgrass prairie. *Soil Biology & Biochemistry* 9: 113-118.

Elliott, E.T. and D.C. Coleman. 1988. Let the Soil Work for Us. *Ecological Bulletins* (Copenhagen) 39: 23-32.

Elliott, E. T., D. C. Coleman, and C. V. Cole. 1979. The influence of amoebae on the uptake of nitrogen by plants in gnotobiotic soil, p. 221-229. In J. L. Harley and R. S. Russell (Eds.) *The

Soil-root interface. Academic Press, London.

Elliott, E. T., R. V. Anderson, D. C. Coleman, and C. V. Cole. 1980. Habitable pore space and microbial trophic interactions. *Oikos* 35: 327-335.

Ettema, C. H., D. C. Coleman, G. Vellidis, R. Lowrance, and S. L. Rathbun. 1998. Spatiotemporal distributions of bacterivorous nematodes and soil resources in a restored riparian wetland. *Ecology* 79: 2721-2734.

Ettema, C.H., R. Lowrance, and D.C. Coleman. 1999. Riparian soil response to surface nitrogen input: Temporal changes in denitrification, microbial and labile C and N pools, and bacterial and fungal respiration. *Soil Biology & Biochemistry* 31: 1609-1624.

Freckman, D.H. (Ed.) 1982. *The Ecology of Nematodes in Soil Ecosystems.* University of Texas Press, Austin.

French, N. R. (ed.) 1979. *Perspectives in Grassland Ecology.* Results and applications of the US/IBP Grassland Biome Study. Springer-Verlag, New York, Heidelberg.

Golley, F. B. 1993. *History of the ecosystem concept in Ecology*: more than the sum of the parts. Yale University Press, New Haven.

Hansen, R.A. 2000. Diversity in the decomposing landscape. In: *"Invertebrates as Webmasters in Ecosystems."* D.C. Coleman and P.F. Hendrix (Eds.) pp. 203-219. CABI Publishing, Wallingford, U.K.

Hansen, R.A. 1999. Red oak litter promotes a microarthropod functional group that accelerates its decomposition. *Plant & Soil* 209: 37-45.

Hansen, R.A., and D.C. Coleman. 1998. Litter complexity and composition are determinants of the diversity and species composition of oribatid mites (Acari: Oribatida) in litterbags. *Applied Soil Ecology* 9: 17-23.

Hendrix, P. F., R. V. Parmelee, D. A. Crossley, Jr., D. C. Coleman, E. P. Odum and P. Groffman. 1986. Detritus food webs in conventional and no-till agroecosystems. *BioScience* 36: 374-380.

Heneghan, L., D.C. Coleman, X. Zou, D.A. Crossley, Jr., and B.L. Haines. 1998. Soil microarthropod community structure and litter decomposition dynamics: A study of tropical and temperate sites. *Applied Soil Ecology* 9: 33-38.

Heneghan, L., D.C. Coleman, X. Zou, D.A. Crossley, Jr., and B.L. Haines. 1999. Soil microarthropod contributions to decomposition dynamics: tropical - temperate comparisons of a single substrate. *Ecology* 80: 1873-1882.

Hobbie, J. E. 2003. Scientific Accomplishments of the Long Term Ecological Research Program: an introduction. *BioScience* 53: 17-20.

Hooper, D.U., J.M. Dangerfield, L. Brussaard, D. Wall, D. Wardle, and 14 co-authors. 2000. Interactions between above and belowground biodiversity in terrestrial ecosystems: patterns, mechanisms and feedbacks. *BioScience* 50: 1049-1061.

Hu, S., D.C. Coleman, C.R. Carroll, P.F. Hendrix, and M.H. Beare. 1997. Labile soil carbon pools in subtropical forest and agricultural ecosystems as influenced by management practices and vegetation types. *Agriculture, Ecosystems & Environment* 65: 69-78.

Hunt, H. W. 1977. A simulation model for decomposition in grasslands. *Ecology* 58: 469-484.

Hunt, H. W., D. C. Coleman, E. R. Ingham, R. E. Ingham, E. T. Elliott, J. C. Moore, S. L. Rose, C. P. P. Reid, and C. R. Morley. 1987. The detrital food web in a shortgrass prairie. *Biology & Fertility of Soils* 3: 57-68.

Ingham, R. E., J. A. Trofymow, E. R. Ingham, and D. C. Coleman. 1985. Interactions of bacteria, fungi, and their nematode grazers: Effects on nutrient cycling and plant growth. *Ecological Monographs* 55: 119-140.

Knapp, A.K., J.M. Briggs, D.C. Hartnett, and S.L. Collins (Eds.) 1998. *Grassland Dynamics: long-term ecological research in Tallgrass Prairie.* Oxford University Press, New York.

Knoepp, J. D., D. C. Coleman, D. A. Crossley, Jr. and J. Clark. 2000. Biological Indices of Soil Quality: an ecosystem case study of their use. *Forest Ecology & Management* 138: 357-368.

Lamoncha, K.L., and D. A. Crossley, Jr. 1998. Oribatid mite diversity along an elevation gradient in a southeastern Appalachian forest. *Pedobiologia* 42: 43-55.

Linden, D., P.F. Hendrix, D.C. Coleman and P.C.J. van Vliet. 1994. Soil faunal indicators of soil quality. Chapter 6, pp. 91-106. in: J.W. Doran, D.F. Bezdicek and D.C. Coleman (eds) *Defining soil quality for a sustainable environment.* SSSA Special Publication Number 35, Madison, Wisconsin.

Macfadyen, A. 1962. *Animal Ecology: Aims and Methods.* Pitman, London.

Magnuson, J. J. 1990. Long-term ecological research and the invisible present. *BioScience* 40: 495-502.

Magnuson, J. J., T. K. Kratz, T. M. Frost, C. J. Bowser, B. J. Benson, and R. Nero. 1991. Expanding the temporal and spatial scales of divergent ecosystems: roles for LTER in the United States. Pp. 45-70 in: Risser, P. G. (Ed.) *Long-term Ecological Research. An international perspective.* SCOPE 47, John Wiley & Sons, Chichester, U.K.

Maxwell, R.A. and D.C. Coleman. 1995 Seasonal dynamics of nematode and microbial biomass in soils of riparian-zone forests of the southern Appalachians. *Soil Biology & Biochemistry* 27: 79-84.

Mitchell, R., R. A. Mayer, and J. Downhower. 1976. Evaluation of 3 Biome programs. *Science* 192: 859-865.

Moore, J.C., E. L. Berlow, D.C. Coleman, and 14 additional authors. 2004. Detritus, Trophic Dynamics, and Biodiversity. *Ecology Letters* 7: 584-600.

Odum, E.P. 1960. *Fundamentals of Ecology*, 2nd Edn., Saunders, Philadelphia.

Odum, H.T. 1971. *Environment, Power and Society.* Wiley-Interscience. New York.

Pesek, J., S. Brown, K.L. Clancy, D.C. Coleman, R.C. Fluck, and 12 co-authors. 1989. *Alternative Agriculture*; Committee on the role of alternative farming methods in modern production agriculture. *Board on Agriculture, National Research Council. National Academy Press.* Washington, D.C. 448 p.

Petersen, H.A., and M. Luxton. 1982. A comparative analysis of soil fauna populations and their role in decomposition processes. *Oikos* 39: 287-388.

Phillipson, J. (Ed.). 1970. *Methods of study in soil ecology.* Proc. of Symposium. IBP-UNESCO, Paris.

Reuss, J. O., and G. S. Innis. 1977. A grassland nitrogen flow simulation model. *Ecology* 58: 379-388.

Rhoades, C.C., G.E. Eckert, and D.C. Coleman. 1998. Effect of pasture trees on soil nitrogen and organic matter: Implications for tropical montane forest restoration. *Restoration Ecology* 6: 262-270.

Rhoades, C.C., G.E. Eckert, and D.C. Coleman. 2000. Soil carbon differences among forest, agriculture and secondary vegetation in lower montane Ecuador. *Ecological Applications* 10: 497-505.

Robertson, G. P., D. C. Coleman, C. S. Bledsoe, and P. Sollins, Eds. 1999. *Standard Soil Methods for Long-Term Ecological Research,* LTER Network Series. Oxford University Press, New York. 462 p.

Simard, S. 2020. *The Mother Tree.* Simon and Schuster, New York.

Six, J., E.T. Elliott, and K. Paustian. 2000. Soil macroaggregate turnover and microaggregate formation: a mechanism for C sequestration under no-tillage agriculture. *Soil Biology & Biochemistry* 32: 2099-2103.

Six, J., H. Bossuyt, S. Degryze, and K. Denef. 2004. A history of research on the link between (micro) aggregates, soil biota, and soil organic matter dynamics. *Soil & Tillage Research* 79: 7–31

Stout, J. D., A. D. Bawden, and D. C. Coleman. 1984. Rates and pathways of mineral nitrogen transformation in soil from pasture. *Soil Biology & Biochemistry* 16: 127-131.

Tian, G., R. Justicia, D. C. Coleman and C. R. Carroll. 2007. Assessment of soil and plant carbon levels in two ecosystems (woody bamboo and pasture) in montane Ecuador. *Soil Science* 172: 459-468.

Van Dyne, G.M. (Ed.) 1969. *The ecosystem concept in natural resource management.* Academic Press, NY.

van Vliet, P.C.J., M.H. Beare and D.C. Coleman. 1995. Population dynamics and functional roles of Enchytraeidae (Oligochaeta) in hardwood forest and agricultural systems. *Plant & Soil* 170: 199-207.

Vannier, G. 1987. The porosphere as an ecological medium emphasized in Professor Ghilarov's work on soil animal adaptations. *Biology & Fertility of Soils* 3: 39-44.

Walker, T.W., and J.K. Syers. 1976. The fate of phosphorus during pedogenesis. *Geoderma* 15: 1-19.

Wall, D.H. (Ed.). 2004. *Sustaining Biodiversity and Ecosystem Services in soils and sediments.* SCOPE Vol. 64, Island Press, Washington, D.C.

Wall, D.H., R.D. Bardgett, V. Behan-Pelletier, J.E. Herrick, and 5 others (Eds.). 2012. *Soil Ecology and Ecosystem Services.* Oxford University Press, Oxford.

Wang, G. M., D.C. Coleman, D.W. Freckman, M. I. Dyer, S.J. McNaughton, M.A. Acra, and J.D. Goeschl. 1989 Carbon partitioning patterns of mycorrhizal versus non-mycorrhizal plants. Real-time dynamic measurements using 11CO2. *New Phytologist* 112: 489-493.

Whitman, W. B., D. C. Coleman, and W. J. Wiebe. 1998. Perspective. Prokaryotes: The unseen majority. *Proceedings of the National Academy of Sciences* 95: 6578-6581.

Wiegert, R.G., and R.G. Lindeborg 1964. Stem-well method of introducing radioisotopes into plants to study food-chains.

Ecology 45: 406-408.

Wiegert, R. G., D. C. Coleman, and E. P. Odum. 1970. Energetics of the litter-soil subsystem, p. 93-98. In J. Phillipson (Ed.), *Methods of study in soil ecology.* Proc. of Symposium. IBP-UNESCO, Paris.

Witkamp, M. 1971. Soils as components of ecosystems. *Annual Review of Ecology and Systematics* 2: 85-110.

Witkamp, M. and J. van der Drift. 1961. Breakdown of forest litter in relation to environmental factors. *Plant & Soil* 15: 295-311.

Wright, C.J., and D.C. Coleman, 2002. Responses of soil microbial biomass, nematode trophic groups, N-mineralization and litter decomposition to disturbance events in the southern Appalachians. *Soil Biology & Biochemistry* 34: 13-25.

Wolters, V., W.H. Silver, D.E. Bignell, D.C. Coleman, and 14 co-authors. 2000. Effects of Global Changes on above- and below-ground Biodiversity in terrestrial ecosystems: Implications for Ecosystem Functioning. *BioScience* 50: 1089-1098.

Woodmansee, R. G. 1978. Additions and losses of nitrogen in grassland ecosystems. *BioScience* 28: 448-453.

Woodmansee, R.G., J.C. Moore, D. Ojima, L. Richards (Eds.) (2021). *Natural Resource Management Reimagined: using the systems ecology paradigm.* Cambridge University Press, Cambridge.

Yeakley, J.A., B.W. Argo, D.C. Coleman, B.L. Haines, B. D. Kloeppel, J.L. Meyer, W.T. Swank, and S.F. Taylor. 2003. Hillslope nutrient dynamics following upland riparian vegetation disturbance. *Ecosystems* 6: 154-167.

Yeates, G.W. 1981. Soil nematode populations depressed in the presence of earthworms. *Pedobiologia* 22: 191-195.

www.ingramcontent.com/pod-product-compliance
Lightning Source LLC
Chambersburg PA
CBHW040301170426
43193CB00021B/2973